On Solid Ground

Why the Earth Isn't as Controversial as You May Think

DAVID GOLDSMITH

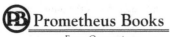

Prometheus Books

Essex, Connecticut

Prometheus Books

An imprint of Globe Pequot, the trade division of
The Rowman & Littlefield Publishing Group, Inc.
4501 Forbes Boulevard, Suite 200, Lanham, Maryland 20706
www.rowman.com

Distributed by NATIONAL BOOK NETWORK

British Library Cataloguing in Publication Information Available

Library of Congress Cataloging-in-Publication Data Available

ISBN 978-1-63388-830-2 (cloth : alk. paper) | ISBN 978-1-63388-831-9 (ebook)

♾™ The paper used in this publication meets the minimum requirements of American
National Standard for Information Sciences—Permanence of Paper
for Printed Library Materials, ANSI/NISO Z39.48-1992.

For Melissa

Contents

Acknowledgments

Like the theory of plate tectonics, this book is a product of history and owes its existence to many people. Thanks to all of my friends, family, and colleagues for their support over the years and for their constant willingness to listen to me while I'm geeking out over a rock or fossil. I would like to recognize a few people individually for their particular roles in the origin of this book. Thanks to Christy Seifert for answering a million questions about how a book gets published and for starting almost every email with "That's great!" Thanks to Stephen Jay Gould for teaching me that the story behind the science is often as interesting as the science itself, and for giving me a literal front-row seat to watch a scientist explain their work in a manner that was both accessible and fascinating. Thanks to Jonathan Kurtz and all of the editors at Prometheus for thinking this was a good idea. And thanks, most of all, to Melissa Goldsmith for being the best reviewer, test audience, cheerleader, friend, and partner a person could ever hope for. I especially couldn't have done it without you.

Introduction

Only about 9 percent of Americans have ever taken an earth science class. This number is particularly jarring considering that 100 percent of Americans live on earth. The fact that so few people ever learn about how the earth works might explain why so much of geology has become controversial. There used to be a rule of thumb that the three topics never to be mentioned in polite conversation were sex, religion, and politics. It might now be time to add geology to that list because it has somehow become a fraught and volatile topic. Hardly anyone argues with their family about atomic theory or general relativity. It's easy, however, to start a fight at many Thanksgiving tables by bringing up climate change, evolution, the age of the earth, and even whether or not the earth is round.

Each chapter in this book is based on a single geologic controversy (such as "the earth is round") and does two things: explains why geologists are so sure about the right answer to that controversy and analyzes the arguments being used by those still unwilling to accept geologic expertise. As it turns out, geologists have learned quite a bit about our planet over the years.

One of the first things scientists needed to realize about the earth before there could even be a science of geology is that it is innately understandable. Geologists can ask questions about it and develop reasonable answers to those questions. Through those inquiries, they have learned that there are things about the planet that have not changed for billions of years. Those constants are the topic of the first half of this book. They include basics such as the earth being round; revolving around the sun; and being solid, old, and a product of history. That is to say, one of the constant things about the earth is that it is constantly changing.

Aspects of the earth that change over time are addressed in the second half of the book. Understanding our changing past requires different methods than demonstrating the constant aspects of the planet, so the second half of the book begins with a description of how geological methods and philosophies have changed over time. From there, it explores how the earth itself has changed, including the evolution of life, moving of continents, and changing of climate patterns. The one thing that has changed most, at least over the past several centuries, is our understanding of the earth as a whole.

Many of the discoveries that have improved our understanding of the earth have been met with resistance. Sometimes that resistance has come from genuine skeptics, which is a good thing for science. Skepticism can be a check that a new theory explains our observations better than a previous theory. Only after a new theory has passed that skeptical test should it be accepted as our best explanation of a phenomenon.

Sometimes resistance to a theory comes about because people are honestly mistaken about some aspect of science. It may be that they are misunderstanding something about the theory itself, or that the theory contradicts something else that they sincerely believe. This resistance that comes from confusion, like resistance that comes from skepticism, can be helpful to science. Science is a collaborative endeavor, and therefore requires clear lines of communication. When scientists encounter honest confusion about what their theories mean, it helps them retool their communication. Unfortunately, not all confusion about scientific theories arises honestly. Sometimes confusion has help.

Sometimes resistance to an idea comes from bad actors looking to sow doubt, not based on the content of a theory, but in pursuit of their own agendas. When a new discovery threatens power structures or income streams, resistance to that theory is inevitable. Consider the link between tobacco use and lung cancer as an example. The link was first discovered in the 1940s, and was well-publicized by the 1950s, but was still a minority opinion among healthcare professionals in the 1960s and is still considered controversial by some people today.[1] This pattern of confusion was not due to uncompelling data or the complexity of the idea. It was intentionally fostered by the tobacco industry.[2] Bad actors aren't always trying to preserve entrenched power. Sometimes they see resistance to a discovery as an opportunity to build a following, or just to bilk the gullible.

*   *   *

This is a book about the earth, but it is not meant to be a comprehensive geology textbook. The topics in this book were chosen specifically because people have argued about them in the past, and in many cases, still argue about them today. So, this is also a book about arguments and how to evaluate their merit.

Strictly speaking, there are two things needed to make a good argument. The first is valid premises. The statements that form the foundation of an argument must be verifiable facts. Geology, and science in general, has several tools at its disposal to ensure that its theories are built on a foundation of solid facts. These include direct observation, precise description, and an ever-increasing arsenal of scientific instruments. Geologists have observed and described the earth at every scale from the microscopic examination of mineral crystals to satellite images of entire continents.

The second requirement for a good argument is sound logic. The conclusions should be the natural consequences of the premises. This is harder than it sounds. In fact, Aristotle, who was one of the first people to concern themselves with the rules for good arguments,[3] was also one of the first to enumerate the traps that people can fall into when trying to make a good argument.[4] These rhetorical missteps have come to be known collectively as "logical fallacies," and while Aristotle enumerated a few of them, today logicians and philosophers recognize hundreds.[5] This book will use the history of geology to illustrate several of these fallacies.

Engaging in legitimate scientific debate requires one more thing beyond a good argument based on facts and sound reasoning. It requires arguing in good faith. This is not about the content or structure of the argument, but about being an honest discussant. Don't knowingly say things that aren't true. This may seem obvious, but depending on the agenda of the person making the argument, sometimes truthfulness is treated as more of a guideline than a hard and fast rule. In some cases, people make bad-faith arguments just so they can say that an argument exists—that not everyone agrees on the right answer to a question or the right solution to a problem. If a person's only goal is to sow uncertainty, then they don't need to go to the trouble of making a good argument. They just need to make an argument.

Another part of debating in good faith is to acknowledge when the other person in the discussion has made a legitimate point. An honest discussant can change their mind. They don't stick to a position in spite of the facts. They revise their position in light of the facts. Scientific theories about the earth have changed considerably over the centuries. That's not due to capriciousness or whimsy. It's because geologists have gained new knowledge and made new observations.

Geology is, first and foremost, a human endeavor. As such, its ideals and best practices have always had to coexist with human nature. The basic rules of scientific debate are simple: Be truthful; pay attention to data; reason well; be persuadable. Follow these rules, and honest discourse should lead to learning. Unfortunately, not everyone can stick to this path. People of good will can stray from it due to pride, a sense of tradition, honest confusion, or misplaced confidence. Bad actors can break these rules willingly in pursuit of their own power, gain, or social agenda. This is a book about the history of geological arguments, the logic behind those arguments, and the consequences of those arguments for our understanding of the earth.

Part I

CONSTANTS

CHAPTER 1

The Earth Is Knowable

By some reckonings, science started on a Wednesday—more precisely, Wednesday, May 28, 585 BCE. According to ancient sources, that was the first time any human being ever successfully predicted a solar eclipse.

The eclipse in question was well-documented after the fact. The historian Herodotus included it in his account of a battle between the Medes and the Lydians, two ancient factions who lived in what is today Turkey and Iran.[1] Herodotus described how the darkening of the sky caused both armies to throw down their weapons in terror and end not just the battle but their long war. The fortuitous coincidence of an eclipse ending a battle makes this skirmish between the Medes and the Lydians the oldest event in the historical record that we can date precisely to the day.[2]

Modern astronomers, with their understanding of the relative motions of the sun, earth, and moon, can predict eclipses centuries in advance. They can also work backward to determine the precise time and place of past eclipses—which is how we can date that ancient battle so precisely. But they can make these impressive predictions and reconstructions only because of how they believe the universe works. And that belief became possible on Wednesday, May 28, 585 BCE.

The man who predicted the eclipse was Thales, a philosopher, mathematician, and astronomer from the Turkish island of Miletus who was born somewhere around 624 BCE.[3] What's remarkable about Thales's prediction isn't the calculation itself. Based on the precision of the data available to him, Thales could easily have been off by weeks; the fact that he predicted the eclipse to the day was undoubtedly a bit lucky. What is remarkable about Thales's prediction is that he made it at all.

In Thales's day, everyone knew what caused an eclipse—angry gods. This wasn't a uniquely Greek opinion. Pretty much anywhere in the world at the

3

time, if there was an eclipse, everybody knew that some god, hero, demon, or other mythic figure was involved. In Thales's Greece, it was Apollo. The details varied from place to place, but the root cause of an eclipse was the same world-wide: someone, somewhere in the universe, was displeased. Of course, it wasn't just eclipses that let us know that the gods were angry. Earthquakes were the fury of Poseidon. Anger Hephaestus and he might make a volcano erupt. The wrath of Zeus was meted out in thunderbolts. All of nature was a manifestation of the gods' moods, and the gods were very moody.

If that's how you believe that nature works, then making any kind of predictions about natural phenomena is impossible. Any natural disaster could occur at any time, because someone out there could be annoying any god at any time. Prediction requires something that mythic explanations for nature prohibit: the underlying causes of natural phenomena to be constant. In simpler terms, it requires believing that the same cause will always lead to the same effect.

It's important to note that Thales had no idea why or how eclipses happened. The idea that they involved the shadow of the moon wouldn't be proposed until almost one hundred years after Thales died. What Thales realized was that eclipses happened with a regular frequency that could be extrapolated forward.[4] If you were to ask him how they occurred, his best answer would probably just be "predictably." It was a radical idea at the time.

Thales's original writings have been lost to history. His accomplishments are preserved for us today through the writings of other philosophers and historians, including Herodotus, Heraclitus, and Democritus. All of them write about his prediction in awe. In the centuries since his death, philosophers from Aristotle to Isaac Asimov have described Thales as the first scientist.[5] They did so because until you move your explanations from the mythical to the logical—from the capricious to the predictable—you can't do science.

Every modern principle of science is based on Thales's assumption that nature is predictable and regular. That is to say, all of science assumes constant natural law. The alternative is a universe of chaos. Even the most basic experiments could only provide provisional results. Does lead float? Not the last time I checked, but nobody knows what the gods may ordain tomorrow.

Once you believe that constant natural laws exist, you can set out to determine what those laws might be. In other words, you can do science. But there are some limits to what prediction and testing can accomplish. Let go of a tennis ball and it will fall toward the ground. That's a testable prediction. You can perform the experiment and see what happens. You can even perform variations on the experiment. You can let go of a tennis ball in your kitchen, or outside, or at night, or next Thursday, or from the top of a ladder to see what happens. But you can't decide today to let go of the tennis ball last week. The past has happened and can no longer be observed, influenced, or experimented on.

One of the most common misconceptions about geology is that it studies the past. But again, studying the past is impossible. You can't perform experiments in the past. You can't observe the past. Our belief that geologists study the past is reflected in the way we describe what they do. We say that geology studies how mountains *formed*, how the continents *moved*, or how rivers *shaped* the landscape. But those processes are still ongoing. Mount Everest is taller today than it was even yesterday. Every eruption of a volcano in Iceland pushes North America a little farther from Europe. The same Colorado River that carved the Grand Canyon over millions of years is still flowing through the canyon, and still eroding away the surrounding rock. The earth is not a completed project. It's a work in progress. Geology studies the present to reconstruct and understand the past.

All sciences are rooted in the axiom that constant rules of nature exist and are discoverable. Geology, however, puts that axiom front and center in its pursuits and gives it a unique name. The *Principle of Uniformity* or *Uniformitarianism* states explicitly that the laws of nature operate today in the same way that they have operated in the past.[6] Geologists need this axiom because the past is gone. To understand what happened in the past, we need to be able to assume that the present is a result of past events and that those past events were not different or unique just because they are past. They are the result of the same processes going on outside our window today.

Thales showed us that our world is a knowable place; but for many people, it is an unknown place. Most Americans have never learned any geology. Many high schools don't offer it. Those that do often call it an elective. Even if you just focus on college-bound high school graduates, fewer than half of them have had any kind of class about how the earth works.[7] That's a shame, not just because the earth is our shared home, but because it is kind of an amazing place.

What's even more of a shame is that there are people who want to exploit our ignorance of geology. The world is home to a broad range of geological doubters and naysayers. They go by many different names—flat-earthers, young-earthers, creationists, climate change deniers, and so forth. They have a range of agendas. Some of them have a financial interest in what they're selling. Some are religiously motivated. Some just like being iconoclastic or difficult. But they all have one fundamental thing in common: they cheat.

If you want to win an argument about what the world is like, there are two fundamental strategies you can use. One option is to have the facts on your side. The world is, after all, an observable place. No amount of rhetoric or nuance is going to change the fact that the earth is round. Or that the earth is old. Or that the earth is not the center of the universe. Since you can't really argue against facts, the alternative is to argue against the existence of facts. This is why you will so often hear the facts of geology dismissed as "just a theory."

Unfortunately, many of the words that scientists use to differentiate their assumptions, observations, and explanations are used differently in science than they are outside of science. As such, it can often be confusing when scientists try to communicate outside their own circle. Worse still, those words can be abused by hucksters and bad actors to undermine science or even the basic premise that the earth is knowable. It is therefore worth defining a few terms.

When scientists use the word *fact*, they are talking about an observation. Lead does not float. That is a fact because it is observable. Measurements are considered a type of observation and are therefore also facts. As of this writing, the seven years from 2014 to 2020 were the hottest years ever measured. We can argue about what to do about that fact, or what caused that fact, but the fact itself won't change. And that's the thing about facts. They rarely change.

What do change from time to time are our explanations for the facts. We call these explanations *theories*. This is where things begin to get confusing. To most people, the word *theory* implies uncertainty or guesswork. Calling something a theory makes it seem like less than a fact. Instead, theories help us understand the facts.

Occasionally, a theory will need revision because scientists have discovered new facts, but how we explain the facts does not change the facts themselves. In the 1980s, geologists discovered evidence that a large extraterrestrial object had struck the earth at about the same time that the dinosaurs went extinct.[8] This caused us to radically rethink our theories about what caused the extinction. It was a period of lively and intense debate over our theories, and yet, sadly, through it all, the underlying fact that dinosaurs are dead did not change.

If a theory sounds tentative and uncertain, then a scientific *law* has just the opposite connotation. Laws are the pronouncements of lawmakers, after all, and will be enforced with rigid penalties if they are violated. But that's not how scientists use the word *law*. A law in science is a way of organizing our observations and predicting the things we haven't yet observed. Generally speaking, laws take the form of equations.

It is possible, then, for a single phenomenon to exist as a fact, a theory, and a law all at once. Let go of a tennis ball and it will fall toward the ground. That is the fact of gravity. Let go of a tennis ball and it will fall toward the ground because objects that are free to move will be moved toward each other by an instantaneous force. That is Newton's particular theory of gravity. Let go of a tennis ball and it will move toward the earth at a predictable rate determined by its mass, the earth's mass, and the distance between them. That is Newton's Law of Universal Gravitational Attraction.

Finally, there are *axioms*. These are the assumptions that we need to make if we are even going to try to understand our universe. There are remarkably few of these, and Thales gave science its first: the laws exist, and they are knowable. It is the axioms that define the boundaries of a discipline.

When you hear someone say that evolution, climate change, or even gravity is "just a theory," they are taking advantage of the fact that the word *theory* has a slightly different meaning outside of science. Referring to Darwinism as "evolutionary theory" does not put it on equal footing with eccentric theories, such as the UFO conspiracy theory or reptiloid theory. To imply otherwise is to commit a type of rhetorical fraud called "persuasive definition," a logical fallacy in which the person making an argument establishes the definition of some key term in a way that biases the discussion in their favor.[9] If science is going to refer to its explanations as theories, and I can convince you that the definition of *theory* implies uncertainty, then convincing you that science is uncertain should be easy. Even if it isn't true.

Geologists have used the axiom that nature's laws stay constant to develop theories that can explain the facts of nature. Along the way, they have discovered that it is not just nature's laws that have remained constant. Quite a bit else about our planet has stayed constant over its long history, including its shape, its size, its relative position in the universe, and its overall composition. The next several chapters are dedicated to those discoveries.

CHAPTER 2

The Earth Is Round

Nobody really knows who first discovered that the earth is round. The shape of the earth is really less of a discovery and more of an observation. Almost anyone on a high enough mountain, or far enough out to sea, can't help but notice it. Trying to ascribe credit for the discovery would be like trying to decide who discovered the moon. It was there for anyone to see if they would just look. And yet, in 2018, two hundred people attended the International Flat Earth Convention in Alberta.[1] That same year, a YouGov survey of ten thousand Americans found that 7 percent either believed the earth to be flat or had doubts about its sphericity.[2] Calling these notions antiquated is insulting to antiquity. Nobody back then actually believed in a flat earth.

The idea that ancient people believed in a flat earth is, ironically, relatively modern. It originated in the 1820s as nothing more than a bit of artistic license in a biography of Christopher Columbus.[3] By the middle of the nineteenth century, however, some people in England were at least pretending to have serious doubts about the shape of the earth. These doubts were almost certainly insincere and inspired more by hopes for financial gain than for legitimate scientific discourse. The fact that flat-earth theory persists at all in an era of space travel and GPS is a testament to the power of a bad-faith argument.

Obviously Round, But How Big?

Some of the oldest surviving scientific writings start from the premise that the earth is round and quickly move on to estimating its size. Eratosthenes was a poet and mathematician born in Cyrene, in what is now Libya, in about 276 BCE.[4] Eratosthenes was a prolific writer on a wide range of subjects. Today he is best remembered for his measurements of the earth's circumference, but in his own time, he was considered a somewhat average scholar. His contemporaries nicknamed him "Pentathlos," a word used to refer to an athlete who regularly comes in second.[5]

Despite having a modest reputation, and apparently some mean friends, Eratosthenes made a name for himself throughout the Mediterranean. He became the personal tutor to the son of Pharaoh Ptolemy III Euergetes. By the time he was thirty, Eratosthenes was made a librarian at the Library at Alexandria.[6]

The Library at Alexandria was possibly the greatest single repository of information in world history until the creation of the internet. Alexandria was a major trading hub in the eastern Mediterranean, and all ships coming to harbor there were searched for scrolls. Any written material that entered Alexandria was confiscated and copied. The copies were returned to the owner. The originals went to the library.

Even before his installation as a librarian, Eratosthenes was fascinated by astronomy and geography. After his installation, he had the resources of information necessary to make great discoveries in those fields. Eratosthenes was among the first to suggest that a year had 365 days in it. He even suggested that every fourth year should be a bit longer.[7] Perhaps Eratosthenes's most noteworthy work at Alexandria was his calculation of the earth's circumference, which began with a simple observation.

Eratosthenes claimed to notice that when he visited the city of Syene (now Aswan in Egypt) on the summer solstice, there were no shadows on the walls of the wells at midday. Strictly speaking, this isn't completely accurate. Syene is slightly outside of the tropics, so there would always be a little bit of shadow. Nevertheless, Eratosthenes reasoned that the sun must therefore be directly overhead on that day so that sunbeams would go straight down the well. Furthermore, Eratosthenes noted that in Alexandria, there was never a day when the wells cast no shadows. He realized that this difference in shadows came about because of sunlight hitting the earth at different angles at different latitudes.[8]

Even in Eratosthenes's day, people knew that the sun is very large and very far away. To give you a sense of just how large the sun is, imagine that you were trying to build a model of the solar system using a basketball as the earth. A standard basketball is a little less than ten inches in diameter. At that scale, the sun would be a little more than ninety feet in diameter. In other words, if the earth were about the size of a basketball, the sun would be about the size of a basketball court. Perhaps more astoundingly, if the earth were the size of a basketball, and the sun were the size of a basketball court, the distance between them would be just a little shy of two miles.

Since the sun is so large, and so far away, beams of sunlight are essentially parallel to one another when they strike the surface of the earth. The assumption that sunbeams are parallel to one another is very important. Because if sunbeams are all parallel to one another, then shadows should be as well. Objects with identical heights, sitting on the same flat surface, should have sunbeams hitting them at the exact same angles, and should therefore cast identical shadows. If they don't cast identical shadows, then one of your assumptions is wrong. With his well observations, Eratosthenes could see the consequences of the earth's roundness. Since shadows weren't constant across the surface of the earth, then the angle between the earth's surface and the sun wasn't constant. (See figure 2.1.) Eratosthenes had discovered a physical consequence of the earth's roundness. He realized that if he could measure how different shadows were from place to place, he could measure how round the earth was. His method was remarkably simple. The only instrument he really needed to measure the entire circumference of the earth was a stick. By measuring the length of a stick's shadow when it was held perpendicular to the ground in Alexandria and Syene on the summer solstice, he could measure the difference in the angle between the surface of the earth and the surface of the sun. Since the sun's rays are parallel to one another, any difference in angle between the sun's surface and the earth's surface must be due to the curvature of the earth. Eratosthenes measured that difference in angle as 7.2 degrees.[9]

As a librarian at Alexandria, Eratosthenes had access to all geographic surveys of Egypt and was, therefore, able to establish the precise north-south distance between Syene and Alexandria. According to the surveys, that distance was approximately five thousand stadia. Historians argue a bit over exactly how long a stadion (singular of *stadia*) was. Several different ancient sources use the term, and unfortunately, they each seem to be describing a slightly different length. If Eratosthenes was using a standard Greek stadion, then a stadion is about 0.115

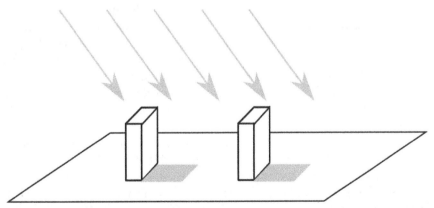

Figure 2.1. If the surface of the earth were flat and sunbeams struck the surface of the earth parallel to one another, then identical objects would have identical shadows.

miles. If he was using Egyptian measurements, it would have been closer to 0.095 miles.[10] Other sources claim that a stadion is simply one-tenth of a mile.[11] For simplicity's sake, since they are all relatively close to one another, and since we can't be sure, we can just average all of these values and say that a stadion is roughly one-tenth of a mile.

Since those five thousand stadia are 7.2 degrees apart, and since 7.2 degrees is one-fiftieth of the way around a circle, Eratosthenes calculated the earth's circumference to be 250,000 stadia, or 25,000 miles. For comparison's sake, the currently accepted value is 24,901 miles.[12] Using little more than a stick and some library research, Eratosthenes attempted to measure the circumference of the entire globe more than two thousand years ago and was astonishingly close to modern estimates. He was only off by about the size of Belgium.

Eratosthenes was not the only ancient Greek who attempted to measure the earth. Posidonius of Apameia was a Greek philosopher born around 135 BCE, about sixty years after Eratosthenes died, in what is today Syria. Posidonius spent his early years traveling extensively throughout the Mediterranean world, as far north as modern France, and as far south as modern Egypt, before eventually settling down to teach in Rhodes. His travels helped him to make observations about how aspects of nature that seem constant if you stay in one location can vary significantly from place to place.[13]

Posidonius's studies of nature kept coming back to astronomy. His study of tidal variations between Greece and Spain led him to realize that tides are determined by the moon, which in turn got him wondering about the distances between the sun, moon, and earth.[14] Posidonius could use geometry to determine the relative distance between the earth and the sun, but to get an absolute distance, he would need to know the size of the earth itself. Posidonius used a completely different method from Eratosthenes to solve this problem. Eratosthenes had looked down at shadows in a well. Posidonius looked up to the stars.

Alpha Carinae (or as it was known in ancient Greece, *Canopus*) is the brightest star in the southern constellation Argo and one of the brightest stars in the sky overall. Posidonius noticed that when he was at home in Rhodes, Canopus only ever appeared at the horizon. But when he was further south in Alexandria, it appeared higher up in the sky. This difference, he realized, is due to the roundness of the earth.[15]

Strictly speaking, just the existence of different constellations in the northern and southern hemispheres is a challenging observation if you believe in a flat earth. Consider for a moment the constellation Crux, also known as the Southern Cross, a constellation that can be seen from anywhere in the southern hemisphere, but that most people in the northern hemisphere have never seen. Figure 2.2 shows how this works.

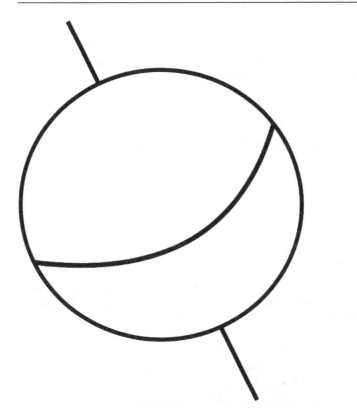

Figure 2.2. The constellation Crux (or the Southern Cross) is located in a part of the sky that is not visible from the northern hemisphere because it is blocked by the curvature of the earth.

Since Crux is in the southern sky, the earth's curvature blocks it from visibility for most people in the northern hemisphere. However, people in the southern hemisphere can see it just fine. In a flat earth model, the phenomenon of different constellations in different hemispheres is trickier to explain. If a constellation is in the sky above one point on a disc-shaped world, it should be visible from every point on the disc. But Posidonius wasn't trying to prove the shape of the earth. That would have been considered old knowledge even back in his BCE days. Posidonius realized that he could use the position of Canopus in the sky to measure the size of the earth. (See figure 2.3.)

When viewed from Rhodes, Canopus sits right at the horizon. When viewed from Alexandria, Canopus sits above the horizon. Posidonius's realization was that geometrically, the difference in angle between the two horizon lines was equal to the angle between the two cities relative to the center of the earth.[16] Since Canopus sat right on the horizon at Rhodes, but 7.5 degrees above the

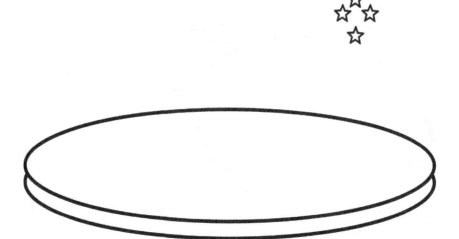

Figure 2.3. If the earth were a flat disc, then every point on its surface should have a line of sight to any constellation in the sky. We would not have different constellations in the northern versus southern hemispheres.

horizon at Alexandria, Alexandria and Rhodes were located approximately 7.5 degrees away from one another relative to the 360 degrees of the circle of the earth—or approximately 1/48th of the way around.[17] (See figure 2.4.)

Posidonius believed that Rhodes was five thousand stadia from Alexandria—exactly as far from Alexandria as Syene was—just in the opposite direction. Posidonius multiplied that five-hundred-mile distance by forty-eight and determined that the circumference of the earth is about twenty-four thousand miles. As a reminder, Eratosthenes said twenty-five thousand miles. Modern science says 24,901 miles.

More than two thousand years ago, two geographers working separately and using completely different methods each measured the size of the earth. Posidonius came within 4 percent of the currently accepted circumference. Eratosthenes was off by less than 1%. But neither of them questioned for even a moment that the earth is round. They were only trying to figure out how round. We have known that the earth is round for millennia.

A Modern Confirmation of Posidonius

The technical term for measurement of the earth's shape is *geodesy*, and it is still an active field of study today. The rough shape of the earth is no longer in doubt, but the fine-scale details are constantly being refined using a very ordinary tech-

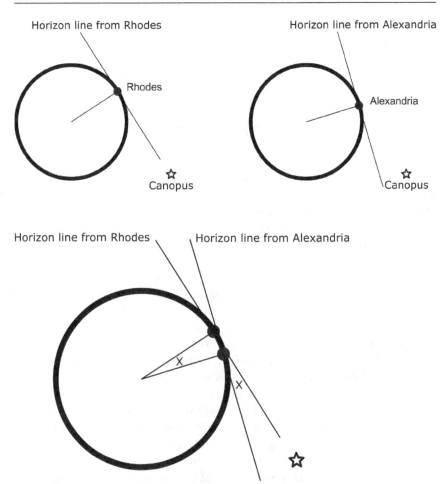

Figure 2.4. The star Canopus sits directly on the horizon line when viewed from Rhodes, but above the horizon when viewed from Alexandria. Posidonius realized that the angle those two horizon lines made relative to the surface of the earth was the same as the angle between Rhodes and Alexandria relative to the center of the earth.

nology. The sensors that geodesists use employ the exact same global positioning system or GPS that exists in your phone, car, or keychain. GPS works not by measuring actual shape, or even distance, but by measuring time.

Compared to other things that we measure, we can measure the passage of time with remarkable precision. In 1967, the International Bureau of Weights and Measures officially defined a second. One second is the amount of time that it takes for an atom of cesium to vibrate 9,192,631,770 times.[18] Contemplate that number for a minute (or 551 billion cesium vibrations, if you prefer). Because it's not 9,000,000,000, and it's not 9,192,600,000. It's 9,192,631,770. That's how precisely we can measure time. Precisely enough to know that if a cesium atom has vibrated 9,192,631,800 times, then 1.000000003 seconds has elapsed.

Because we can measure time so precisely, GPS uses time as a proxy for distance. People do this too, all the time. If you ask someone where they grew up, they are just as likely to reply "about an hour and a half outside New York City," as they are to reply "about eighty miles from New York." Of course, that information is only a little bit helpful. Poughkeepsie, New York; New Haven, Connecticut; and Philadelphia, Pennsylvania, are all "about an hour and a half outside New York City," but all in different directions. However, if you were to ask again, and get the additional information "about two hours from Boston," then we would be getting someplace. To be precise, we'd be getting to New Haven, which is both ninety minutes from New York and two hours from Boston. Knowing your distance from multiple different points allows you to narrow your own location down to a single point. GPS works almost exactly the same way, but instead of using cities, it uses satellites.

Currently, there are several dozen satellites in orbit around the earth that are part of the GPS network. Each one has a high-precision clock on board and is constantly broadcasting the time on that clock to the billionth of a second out into space. A GPS locator listens to the times being broadcast by those satellites. But since the satellites are positioned far apart from one another in orbit, it takes the signal a different amount of time to make the journey from space to the locator. When the signals get to the locator, each is reporting a slightly different time. The locator uses those time differences to determine how far it is from each satellite, and therefore where in space it itself is located. (See figure 2.5.)

Modern geodesists use high-precision GPS data not just to measure the shape of the earth but to monitor changes in its shape as well. The shape, as it turns out, isn't a perfect sphere but an *oblate spheroid*—a sphere that bulges slightly in the middle.[19] This bulge occurs because the earth is spinning rapidly (about a thousand miles per hour at the equator) and is fluid on the inside (more on that in a later chapter). What might be more interesting than the shape, though, is the way the shape changes. Geodesists use their data to track changes in sea level, the motion of continents, the uplift of active volcanoes, and other geologic phenomena that we would never be able to see with our naked eyes.

In a sense, the techniques that we use for geodesy today are based on the same basic principles that Posidonius used millennia ago. Find fixed points in the sky, calculate your distance from those points, and you can determine your position on the surface of the earth.

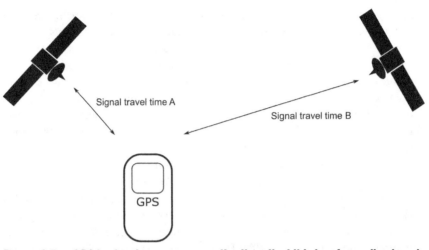

Figure 2.5.   GPS technology measures the time that it takes for radio signals to travel from satellites to a particular locator. By comparing those travel times to one another, the locator can determine its own exact position.

Flat Earth and the Myth of Columbus

If the Ancient Greeks knew almost exactly how round the earth is, then why do so many people believe it was Christopher Columbus who first demonstrated the earth's correct shape? One possible reason might be that the knowledge of the earth's size and shape was simply lost somewhere along the way. A lot of knowledge from ancient Greece has been lost. As the Greek Empire collapsed, and the Roman Empire later followed suit, the Library at Alexandria was repeatedly sacked and burned. Tragically, most of the texts contained in that vast repository (including the original geographic works of Eratosthenes and the original text of Posidonius) were lost to history. But the knowledge itself survived. Eratosthenes and Posidonius were both included in *On the Circular Motions of the Celestial Bodies*, a later anthology by the astronomer Cleomedes that managed to escape destruction. Cleomedes's work was not well-known in Europe immediately after the fall of Rome. Almost nothing of antiquity was known in Europe during the Dark Ages. But these works made it into the modern world thanks to the work of Islamic scholars.

All through Europe's dark ages, Islamic scribes translated ancient works from Greek and Roman into Arabic and disseminated them throughout the Medieval Muslim world. The knowledge that was rescued from Alexandria traveled to Baghdad and eventually to Muslim cities in Europe such as Cordoba, Toledo, and Grenada in Spain.[20] Knowledge of the earth's size and shape was unquestionably available to Columbus. In fact, in 1492, Spain was where most of the ancient world's knowledge was concentrated. Columbus and his contemporaries knew the world was round. What they disagreed about was its size.

To be fair, there was some room for disagreement. Repeatedly translating and transcribing the original texts was like a thousand-year-long game of telephone. Some transcriptions introduced changes that were then faithfully transcribed into later versions and new translations of the text. For example, somewhere around the BCE/CE transition, the Greek geographer Strabo made a few corrections to Posidonius's estimates based on improved knowledge of the distances between Alexandria and Rhodes. One hundred years later, these changes were incorporated into Ptolemy's astronomical treatise *The Almagest*.

In the ninth century, in his work *Elements of Astronomy on the Celestial Motions*,[21] the Arab scholar Al-Farghani tried to make *The Almagest* more accessible by publishing its conclusions about the relative sizes and positions of the sun, moon, and earth without all of Ptolemy's math and derivations. As you may imagine, readers vastly preferred Al-Farghani's math-free explanation to Ptolemy's complex geometry and formulae. And so, in 1492, it was Al-Farghani's words, rather than Ptolemy's math or Posidonius's experiments, that informed the educated people of Spain about the size and shape of the earth.[22]

We know from his diaries that Columbus read Al-Farghani's work. To be more precise, we know that he *mis*read it. Al-Farghani gave the circumference of the earth in miles. Unfortunately, not all miles are the same. Since Al-Farghani was living and working in Baghdad, he gave the earth's circumference in the Arabic miles that he and everyone around him used. Since Columbus was born and raised in Italy, he assumed that the measurement was in the Roman miles that he and everyone around him used. The difference between the two units of measure is substantial. Through this simple mistake, Columbus shaved about four thousand miles off the circumference of the earth.[23]

To compound Columbus's troubles, there was considerable debate in the fifteenth century as to exactly how big Eurasia itself was. The actual distance from the coast of Portugal to the coast of China is about 35 percent of the way around the globe. In the 1490s, most people (based on Ptolemy's estimates) believed the continent to span about 50 percent of the globe. However, some ancient texts suggested it might actually be somewhere between 70 percent or 80 percent of the distance around the earth. Columbus chose to believe those larger estimates for the relative size of Eurasia, and therefore assumed that the distance in the opposite direction was only about 20 percent of the size of the earth.[24]

Through a combination of measurement error, mistranslation of ancient texts, and wishful thinking, Columbus convinced himself that the distance between Spain and Japan was about two thousand nautical miles. The right answer is closer to eleven thousand nautical miles. Put another way, Columbus believed that Japan was only about as far from Spain as Finland is. He had already sailed farther than he expected to when he luckily (for him at least, certainly not for everyone involved) bumped into a previously unknown continent. Had the Americas not been there, he and his entire crew would surely have died from a fatal combination of starvation and hubris.(See figure 2.6.)

Notice, though, that all of Columbus's many compounded mistakes have one thing in common. They are all based on readily available contemporary sources that presumed the earth to be round. Nothing that he, or anyone else at the time, had read would have even suggested the earth was flat. So where did the myth of Columbus versus the flat-earthers come from? Like so many famous stories, it came from a famous storyteller.

Washington Irving was the creator of many of our earliest American legends. He gave us the Headless Horseman of Sleepy Hollow and the oversleeping Rip Van Winkle. While Irving is mostly known as a purveyor of fiction, he did occasionally try his hand at historical writing as well. Unfortunately, when he did so, he often had trouble leaving his penchant for mythmaking behind.

In the 1820s, during a time when he was living abroad in Europe, Irving was invited to Madrid to survey a trove of recently discovered manuscripts concerning Spain's exploration and colonization of the Americas.[25] Several of these documents specifically pertained to Columbus's time in the court of Ferdinand

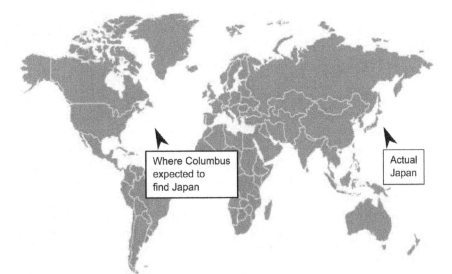

Figure 2.6. Christopher Columbus believed that Eurasia took up 80 percent of a much smaller earth, and that therefore the ocean voyage west to Japan would only be about two thousand miles.

and Isabella. Irving drew on this source material to write what he considered to be his first purely historical book, *A History of the Life and Voyages of Christopher Columbus*, in 1828.

The documents that Irving used for inspiration were nowhere near as detailed as the book that he wanted to write from them. So, where the historical record failed him, Irving employed his imagination. The result is Irving almost single-handedly creating the myth of Columbus's triumph over medieval flat-earthers. He wrote:

> [Court astronomers] observed, that in the Psalms, the heavens are said to be extended like a hide; that is, according to commentators, the curtain, or covering of a tent, which among the pastoral nations was formed of the hides of animals; and that St. Paul, in his epistle to the Hebrews, compares the heavens to a tabernacle, or tent, extended over the earth, which they thence inferred must be flat. Columbus, who was a devoutly religious man, found that he was in danger of being convicted, not merely of error, but of heterodoxy.[26]

This passage is almost undoubtedly a work of complete fiction. Once again, if there was one place on earth in 1492 where educated people knew not just the shape but also the size of the earth, it was (up-until-recently) Moorish Spain.[27]

Nobody really believes that a Headless Horseman haunts the bridges of Sleepy Hollow. Nobody sincerely worries that they might inadvertently oversleep by decades as Rip Van Winkle is alleged to have done. Yet American schoolbooks started including the myth of Columbus versus the flat-earthers as early as 1860.[28] Why did so many people latch on to this one particular figment of Irving's imagination and believe it to be true?

The early nineteenth century saw the transition of geology from a hobby to a science. Prior to about 1800, geology was almost purely a study of rocks and minerals—a pastime more akin to stamp collecting than to physics. By the 1830s, it had become a study of how the earth came to be in its present condition. As theories of the workings and history of the earth proliferated, there was some understandable pushback from fields whose traditional turf geology was now encroaching on. This pushback came from some fields of science (see chapter 5 for an example from physics) but also from a small number of religious scholars concerned with a literal interpretation of the Bible. Geologists responded to this pushback aggressively, perhaps too aggressively. They painted those who disagreed with their new findings as being committed to a medieval worldview and then portrayed medieval scholars as being so scripturally indoctrinated that they believed in a flat earth.[29] It was a convenient shorthand for their opponents' alleged ignorance, and the attack stuck.

There is actually very little in the Bible that definitively weighs in on the shape of the earth. Yes, the Psalms and the Epistle to the Hebrews do, in fact, describe the sky as stretched out over the earth like a tent. But that's a metaphor. In a 2014 episode of the docu-series *Cosmos*, astronomer Neil DeGrasse Tyson describes life on earth as having "awakened on this tiny world beneath a blanket of stars."[30] A blanket of stars is a metaphor almost identical in structure to the tent of the sky mentioned in the Psalms, but nobody could credibly accuse the director of the Hayden Planetarium of pushing the flat-earth agenda.

Samuel Birley Rowbotham: Champion of the Flat Earth

Nobody knows who discovered that the earth is round, but we do know the first person to try to empirically prove that it is flat. That distinction goes to the British inventor Samuel Birley Rowbotham, who lived surprisingly recently (1816–1884). Rowbotham is a complicated character. Depending on the attitudes of their authors, biographies of Rowbotham describe him as everything from a visionary to a quack. Most of our information about him comes from his own writings, which are absurdly self-aggrandizing. In one passage of his memoirs, he reminisces about ditching school in protest, incensed that he was being brainwashed with Newtonian Physics instead of proper Biblical cosmogeny.[31] He was seven at the time of this alleged act of rebellion.

Rowbotham might best be described as a serial charlatan. He seemed to spend his life having a deep-seated curiosity about whatever trendy topic might bring in money next. At various points in his life, he invented railway cars that were never built, ran a socialist utopia, brewed phosphate sodas to compete (unsuccessfully) in the emerging soft drink market, and published books and pamphlets that he claimed demonstrated the flatness of the earth.[32] Some sources allege that he spent part of his later years selling patent medicines under an assumed name and living in a lavish house with an underage wife.[33]

Rowbotham published his books and pamphlets under the pseudonym of Parallax. In retrospect, it seems surprising that a socialist flat-earther with a pen name like an Avengers villain could possibly have been the least bit influential. But Rowbotham's charisma kept people intrigued in his ideas, and his ideas (especially regarding the shape of the earth) were inherently appealing because they were easy to understand.

Rowbotham's geodetic experiments were based on two simple truths of Euclidean geometry: Parallel lines never meet, and a tangent line touches a circle at only one point. Consequently, since light travels through the atmosphere in a straight line, we should have only a limited range of vision on a spherical world.

Figure 2.7 is from Rowbotham's book *Earth Not a Globe*, and is very much not to scale.[34] Imagine a person standing at point T and looking off in the distance toward point N. On a spherical earth, the world would curve increasingly away from their line of sight. Rowbotham calculated that a mile from point T, the surface of the earth would be eight inches lower than that observer. By two miles away, it would be thirty-two inches lower. If the observer had a clear enough view to be able to see three miles away, they would actually be looking six feet (seventy-two inches) off the surface of the earth.

In figure 2.8, Rowbotham shows the consequences of this geometric fact. If an observer were to look at a series of objects of equal height receding further and further into the distance, that observer should be able to see the consequences of the earth's roundness. The further away each object was, the shorter it would appear, until eventually the objects disappeared from view, hidden behind the curvature of the earth. The only way the objects could all stay in view would be if the earth beneath them were a flat surface, parallel to the observer's line of sight.[35]

For Rowbotham to turn these thought experiments into reality, he would need a stretch of countryside not only that provided miles of visibility but also that he could be assured was reasonably flat. Fortunately, he lived in exactly such a place. In 1837, Rowbotham took a position as secretary of the Manea Fen Colony, a utopian socialist commune located in Cambridgeshire along the Old Bedford River.[36]

It is entirely possible that Rowbotham took the position at Manea Fen just so he could use the river for his experiments. The Manea Fen utopia was intended as a place where people lived according to the rules of Robert Owen, a philanthropist and anti-poverty reformer. But Rowbotham was more likely attracted to Manea Fen for reasons of convenience and comfort than philanthropy. The people that Rowbotham recruited into the community seemed to

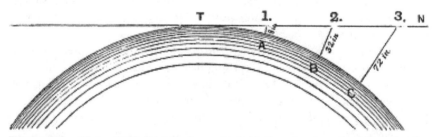

Figure 2.7. Since light travels in a straight line, on a curved earth, the farther you look out toward the horizon, the higher up above the surface of the earth you would be able to see. Beyond a certain distance, the curvature of the earth would block your ability to see objects on the surface. *Courtesy of sacred-texts.com.*

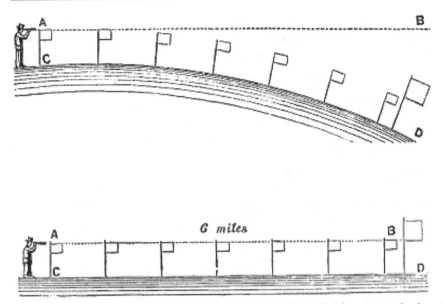

Figure 2.8. On a curved earth, objects of equal height arranged at a series of increasing distances from the observer should look progressively shorter as the earth's surface curves away from the observer. If, on the other hand, their tops appear to line up with each other in a parallel line, that would be evidence of a flat earth. *Courtesy of sacred-texts.com.*

have very little interest in political reform. Reports from the early days of Manea Fen draw far more attention to the party atmosphere than any kind of actual good works being done there.[37]

Whether it was Owenist social reform, scientific curiosity, or something else that brought Rowbotham to Manea Fen, he could scarcely have found a better place to carry out his experiments. The Old Bedford River had recently been turned into a barge canal, resulting in a stretch of perfectly straight, perfectly still water. Here, Rowbotham could look out over a flat stretch of terrain that went on for six miles without so much as a tree to obscure his view. Based on his calculations, an object six miles away on a round earth should also be twenty-four feet below a careful observer, and beginning in 1838, Rowbotham carried out a series of experiments to demonstrate that this was not the case. In each of his experiments, Rowbotham reported that an object six miles downstream appeared no higher or lower in elevation than one on the canal banks.[38]

Some of the experiments that Rowbotham performed were relatively simple ones similar to that illustrated above. He did indeed put a series of flags of equal height on rafts at one mile intervals to see if their tops lined up when viewed from the canal bank. However, some of his experiments were more sophisticated and instrument-based. Good-quality surveying equipment had been invented and put into use in Britain approximately fifty years earlier, and Rowbotham made use of it. In one experiment, he placed a theodolite, essentially a telescope that can be mounted at any precisely measured angle to the ground, midway between two canal bridges to demonstrate that the height of marks on the bridges did not change with distance.[39]

You wouldn't think that there would be much money in the shape of the earth, but Rowbotham's observations were lucrative because they were sensational. The mysterious Parallax filled lecture halls with paying customers and sold books all while irritating the scientific establishment. Even after his death, people continued to promote his work through the Universal Zetetic Society, one of the first modern flat-earth societies, and their self-published periodical, the snappily named *Earth (Not a Globe) Review*. It is understandable why Rowbotham's work caught on as it did. At first glance, his experiments seem like perfectly reasonable ways to demonstrate the flatness of canal water, and therefore the earth, over the space of six miles. What's more, the error in his method is subtle and requires a little knowledge of physics to fully grasp.

Rowbotham's experiments had one central flaw to them. They ignored the fact that the earth has an atmosphere. This could have been an honest mistake. People describing the earth leave the atmosphere out all the time. Even geology textbooks often leave it out of their diagrams of the earth. They begin by listing the rocky crust as our planet's outermost layer, and then only describe the atmosphere in later chapters devoted to weather and climate.

We only tend to think about air at those unfortunate times when we can't get enough of it. Yet we spend our lives completely surrounded by it. As a result, the only way that light arrives at our eyes is by traveling through air. Unless you have been to outer space, you have never seen light unfiltered by the atmosphere. Occasionally, we are reminded that the air is there when we can see the consequences of the interaction between air and light. One example is a heat mirage, in which the air over a hot patch of roadway seems to shimmer—sometimes so much that it makes the road look wet. This is an example of a physical phenomenon called refraction.

Refraction occurs when light rays get bent as they pass through materials with different physical properties. Most typically it occurs when light moves between materials with different densities. It's the reason that the world seems distorted when you try to look at it through a glass of water. As the light passes from air to glass to water to glass to air again, it gets repeatedly bent, and the world behind the glass looks twisted in turn.

Refraction can also occur within the same material if the properties of that material change, either abruptly or gradually. It is this type of refraction that skewed Rowbotham's results. Air near the ground is denser than air higher up. This is well known to any mountain climber, but even in the tens of feet nearest to sea level, there are small but measurable differences. One factor that can increase these differences is the presence of water. Air just above the surface of a body of water is more humid, and therefore even denser than air higher up above the water.

Rowbotham's experiments produced the results that they did not because the earth is flat, but because the path of light over the earth, especially near a body of water like Bedford Canal, is curved. That curve is very slight and over larger distances stops having a noticeable effect, but over the few miles of the Bedford Canal experiments, it was enough to make the earth seem flat. If Rowbotham was unaware of the phenomenon of atmospheric refraction, this might be a reasonable, honest mistake to make. Such ignorance, however, seems unlikely. The phenomenon was well known to astronomers and physicists in Rowbotham's day, having been described by ancient Greeks and Egyptians more than two thousand years prior.[40]

Keeping Faith on a Flat Earth

Rowbotham turned his lectures and pamphlets on the flatness of the earth into a small but lucrative franchise. His readers and converts were writing books and converting followers of their own. One of these flat-earth acolytes was John Hampden, whose healthy inheritance from his late father allowed him a life free from work that he could devote to theological rabble-rousing. To use a modern term, he was an early anti-science troll. Hampden wrote a series of pamphlets called *The Popularity of Error, and the Unpopularity of Truth*, in which he enthusiastically proclaimed Rowbotham's truth over the error of science.[41]

In January of 1870, Hampden decided to put his money where his mouth was and placed an ad in the journal *Scientific Opinion* offering to wager up to £500 against anyone who could prove the earth to be anything other than flat. This ad caught the attention of the biologist Alfred Russel Wallace, who saw it as easy money. Wallace repeated Rowbotham's experiments at the Bedford Canal using equipment that corrected for atmospheric refraction. Unsurprisingly, Wallace's improved experiments clearly demonstrated the curvature of the earth. Equally unsurprisingly, he lost the bet anyway.[42]

There is a rule among gamblers when it comes to proposition bets: Don't take them. If somebody walks up to you in a bar and bets you $5 they can do something that sounds ridiculous like balance an egg on its end or pick up a pint glass with a straw, it's because they already know a trick by which they can do it. Accepting the wager isn't as much placing a bet as it is paying for a show. Wallace should have known going in that Hampden would never have offered such a bet if he had any intention of ever paying it off.

When Wallace and Hampden met at the Bedford Canal to settle their bet, Hampden brought his own referee, a fellow flat-earther named William Carpenter. Each time Wallace performed an experiment to demonstrate the earth's curvature, Carpenter declared the experiment flawed in some way. When Wallace corrected the alleged flaw, and the experiment still demonstrated a curved earth, Carpenter alleged a new problem. No matter how painstakingly Wallace tried to satisfy Hampden and Carpenter's experimental guidelines, any result that showed a round earth was declared obviously flawed because it had shown a round earth.[43]

The reasoning that Hampden and Carpenter used to deny the earth's shape in the face of Wallace's repeated experiments is the same used by modern flat-earthers. It is a logical fallacy called argument from ignorance. The structure of the argument is simple. I believe my position because there is no evidence to the contrary. Every time you show me evidence to the contrary, I dismiss it because it is contrary to my position. No matter what evidence of the earth's roundness you show a flat-earther, they deny the evidence and then continue to proclaim that there is no evidence that the earth is round.

> Why do you think the earth is flat?
> Because there is no evidence that the earth is round.
>
> What about people who have circumnavigated the globe?
> They are clearly using flawed maps.
>
> The fact that other planets are round?
> Earth is not like other planets.
>
> But you can see the earth's shadow on the moon during a lunar eclipse!
> You're making assumptions about how eclipses happen.
>
> Pictures of the earth from the moon?
> You don't really believe we've gone to the moon, do you?
>
> I give up!
> That's because you know there's no evidence that the earth is round.

Flat-earthers are free to use all the flawed logic they want to, but no amount of circular reasoning will change the circularity of the earth. We can see the earth is round by looking at the features on its surface and by mapping the stars in the sky. We can use those same stars to measure the circumference of the earth, and we can verify those measurements with either the shadows on the ground or the same GPS technology we trust to get us where we need to go. Even Christopher Columbus, who could have benefitted from a GPS unit himself, only verified what other people had known for centuries: Earth is round and big.

CHAPTER 3

The Earth Goes around the Sun

There is perhaps no more treasured optical illusion than a sunrise. We have known for centuries that the sun doesn't actually move around the earth, and yet the sun continues to rise in the lyrics of Cole Porter, the poetry of Percy Shelley, and the paintings of Bob Ross. This is probably because it is so easy to just believe the evidence of our senses: The sun definitely looks like it's moving, and we definitely don't feel like we are moving. It's also very hard to shake the yoke of tradition. If you spend your life talking about the sun rising, setting, or moving behind a cloud, that reinforces a model in your mind. So it is perhaps unsurprising that replacing the model of a moving sun with one of a moving earth was a hard-fought battle.

One of the longest-lived theories in the history of science was Ptolemy's model of the universe with the earth at the center. It was conventional wisdom for nearly fifteen hundred years. Trying to overturn that model landed Galileo in front of the Inquisition not once but twice and made him one of the great heroes in the history of science. But at the time when he was fighting this fight, it was not really a foregone conclusion that he was right. In fact, when it comes to Galileo's second inquisition in particular, he might have even had it coming.

An Unmoving Earth as the Center of the Universe

I can remember the first time I was ever skeptical about a geologic theory. I was about six years old, and my grandfather had just told me that the earth spins. He had told me things before that I hadn't believed, and rightly so. (My face never did freeze that way.) But this was different. This story didn't seem like a ploy to get me to be quiet or to stop misbehaving. And he seemed completely serious about it. Still, I was unconvinced.

Part of the problem was that I didn't really understand his model. I imagined the ground somehow moving while everything on it remained in place. I patiently explained to my grandfather that I knew he was wrong because I had recently dug a big hole in the backyard, and it was still in the same place. He found my rebuttal hilarious but unconvincing. But then the topic changed. Where exactly was this hole?

I was not alone in my skepticism. According to a 2020 report by the National Science Foundation, only 72 percent of Americans are aware that the earth goes around the sun.[1] Not only is the idea that the earth spins like a top strange, but it also seems to contradict our daily experience. Science is, in a large part, an exercise in explaining our observations, and we tend to trust our observations. Therefore, when our scientific theories contradict what our senses tell us, it can be difficult to reconcile the two. We can see the sun, moon, and stars all move across the sky. If we were really moving, you would think we would feel it. Add in the fact that while we spin we are also moving around the sun—hurtling through space at supersonic speeds—and a little skepticism is understandable.

The idea that the earth moves around a sun that sits at the center of the solar system wasn't developed by Galileo, Copernicus, or any of the other names we often associate with the model. It is actually much older than that. As early as the third century BCE, Archimedes was describing a controversy over whether the earth was the center of the universe or somewhere off-center.[2] That controversy raged on and off for almost four hundred years before being settled, at least for a while.

Somewhere around the middle of the second century, the Egyptian astronomer Claudius Ptolemaeus (better known today as Ptolemy) wrote *The Almagest*, a book about the structure and workings of the universe. Its title comes from an Arabic phrase meaning "the greatest." That wasn't the original title. It would be bold, bordering on arrogant, to just declare your new book the greatest right out of the gate. Ptolemy originally called it the much more mundane *The Mathematical Composition*. Over years of transcription and translation, *The Mathematical Composition* became *The Great Composition*, and then eventually *The Greatest Composition*, and then finally just *The Greatest*.[3] Ptolemy may not have intended his work to bear such a grandiose title, but the book most definitely earned it. It was the definitive reference on astronomy and cosmology for over a thousand years.[4] And the model of the universe that it presents still bears his name.

The Ptolemaic universe is an intricate and elegant place. The earth is at the center of the universe and remains motionless. The sun and moon orbit the earth in perfect circles and are in constant motion. A little further out are the planets. Their orbits aren't simple circles. Instead, each planet is embedded off-center in a circle called an epicycle. It is the epicycle that moves in a circle around the earth, spinning as it goes. This explains why planets generally follow the same path as the sun and moon, but sometimes appear to slow down, speed up, or even go backward in the sky relative to everything else. The outermost part of the universe is a final crystalline sphere in which the stars are embedded. This model of circles within circles was not original to Ptolemy. In fact, it can be found in Ancient Greek astronomical texts going back five hundred years before Ptolemy's birth.[5] But it was *The Almagest* that secured the model's prominence. (Figures 3.1 and 3.2 depict the universe according to Ptolemy.)

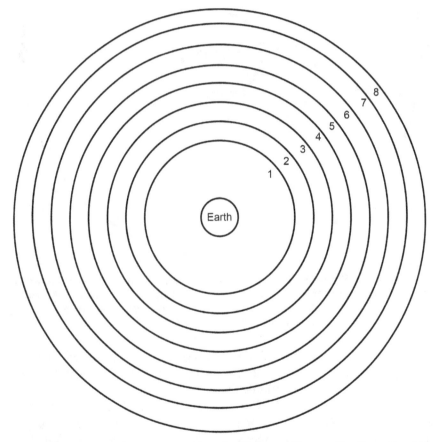

Figure 3.1. The universe according to Ptolemy. The earth sits unmoving at the center of the universe and is orbited at increasing distances by (1) the moon; (2) the epicycle of Mercury; (3) the epicycle of Venus; (4) the sun; (5) the epicycle of Mars; (6) the epicycle of Jupiter; (7) the epicycle of Saturn; (8) the stars.

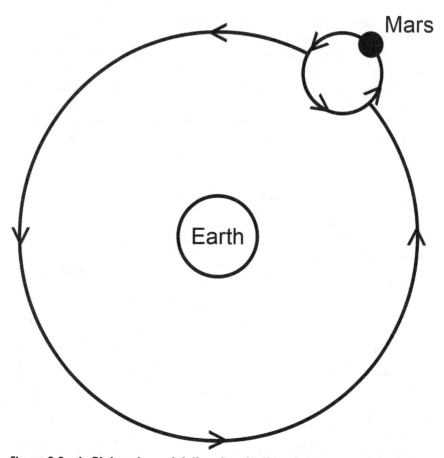

Figure 3.2.   In Ptolemy's model, the planets did not orbit the earth directly. Instead, each planet was embedded in an epicycle that spun as it orbited the earth. This explained why, sometimes, the planets seem to be moving backward relative to other objects in the sky.

Ptolemy's geocentric model of the universe has fallen out of favor (at least with 72 percent of Americans), and it is often lumped together with flat earth theory as a shorthand for scientific illiteracy, but there is actually a lot to like about it. Intricacy and elegance are nice qualities in a scientific model. But Ptolemy's model has something else going for it: It works. It successfully organizes and explains our observations and even makes predictions going forward. After all, you can't be the premier reference book on astronomy for over a millennium if Venus isn't where you say it is.

To clarify, *The Almagest* predicts where in the sky you should look if you want to see Venus. It is way off in terms of how far away Venus actually is. But that wasn't a concern for Ptolemy or any of the countless astronomers who followed in his immediate footsteps. NASA would never be able to land anything on Venus using *The Almagest* as a roadmap. But *The Almagest* did have all the practical information that an astronomer, astrologer, sailor, mathematician, or anyone living prior to 1500 CE could possibly have wanted.

The fact that *The Almagest* was so useful for so long raises some interesting questions. What should we do when a scientific model is useful but also happens to be wrong? For that matter, what does it mean for a model to be right or wrong in the first place? When we judge a statement as true or false, what are we actually saying about that statement?

Typically, when we say a statement is true, we are saying that the statement accurately reflects the way things are. If someone tells me it is sunny out, I can check the truth of that claim by looking out the window. If a claim matches the actual state of things, it is said to have correspondence truth.[6] But there are other ways to test truth.

If someone tells me that it is sunny out, and I am in the Bahamas in July, I probably won't bother to look out the window before judging the truth of that claim. I may just assume it is true because it is consistent with other things that I know to be true about the climate in the Bahamas—it is often sunny there in July. This is coherence truth.[7]

If I am in the Bahamas in July and I want to know if it's lunchtime yet, I can use the fact that it's a sunny day to make that decision. Specifically, I can see how far the sun has moved across the sky and get a rough idea of how close to midday it is. It doesn't matter that I'm using a flawed model. The sun isn't actually moving across the sky. The earth is rotating. What's important is that my model allowed me to estimate the time. This gives my model a kind of truth. A statement has pragmatic truth if believing it is useful in some way.[8]

At the time of its initial writing, Ptolemy's model of the universe had coherence truth. It was perfectly consistent with every assumption we had about how nature worked. It also had pragmatic truth. *The Almagest* was a useful tool for navigation, predicting eclipses, and other astronomical tasks. The geocentric universe had coherent truth and pragmatic truth going for it. It would take nearly a thousand years before anyone started to doubt its correspondence truth.

Moving the Earth and Starting a Revolution

If you wanted to pick a starting date for the beginning of the Scientific Revolution, 1543 is a pretty good candidate. That year saw the publication of some of the first anatomical drawings based on dissected cadavers.[9] It produced the first common-language translation of Euclid's *Elements*.[10] And in 1543, Nicolaus

Copernicus published his new theory on the structure of the solar system in *On the Revolutions of the Heavenly Spheres*.[11] In this book, Copernicus suggested the radical idea that maybe it was the earth and not the sun that was moving. Maybe everything in our little corner of the universe was actually moving around the sun instead. Today we call that idea *heliocentricity*.

Copernicus first put his theories to paper in 1514 in a pamphlet called the *Commentariolus*, or "little commentary." Copernicus sketched out his model of the universe in seven postulates, including the claims that the moon is the only thing orbiting the earth, that everything else goes around the sun, and that any apparent motion of the sun or stars is actually due to the revolution of the earth.[12] Copernicus circulated the *Commentariolus* privately among his friends, so his belief that the sun was at the center of the solar system was well known long before *On the Revolution of the Heavenly Spheres* was ever published.

Rumors of Copernicus's beliefs traveled widely throughout Europe and found favor in surprising places. In 1536, he received a letter from Nikolaus von Schönberg, the cardinal of Capua. Cardinal Schönberg had just acquired a new secretary, who had told him of Copernicus's heliocentric ideas.[13] This secretary, incidentally, had previously been in the employ of the recently deceased Pope Clement VII and had also shared Copernicus's views with him—apparently much to the pope's delight. In the letter, Cardinal Schönberg begged Copernicus to finally publish his ideas and, in fact, offered up the services of a scribe and messenger to make publication easier.[14]

If it seems odd that both a cardinal and a pope in the sixteenth century would be delighted by heliocentricity, then you may have fallen victim to a logical fallacy—the fallacy of hasty generalization. This fallacy occurs when you extrapolate a general rule from a few instances that might not accurately represent every member, or even typical members, of a group. One way this fallacy manifests itself is in the perception that there is a never-ending war between science and religion. There isn't.

Science and *religion* are both very broad terms that encompass a wide range of practices and beliefs. And in most of the particulars, there is no conflict. There is, for example, no conflict between Judaism and plate tectonics. Not only is quantum theory perfectly compatible with Buddhism, but also the Dalai Lama has even written a book on how interesting he thinks it is.[15] Undeniably, there have been some spectacular clashes throughout the history of science between specific scientific theories and individual religious leaders (as we will see later in this chapter), but it is perfectly possible to be a devout person and a curious scientist. Pope Francis once worked as a chemist in a food lab,[16] and Copernicus himself was a canon of the Catholic Church—an expert in liturgical law.[17]

Despite being a canon himself, having a cardinal as a patron, and having heard rumors of a pope unfazed by heliocentricity, Copernicus was still wary of publishing his ideas. Some of this stemmed from lingering worries that he would be accused of heresy. However, Copernicus was also concerned about the reception of his model in general. It did a much better job of explaining and describing the solar system than Ptolemy did, but it still wasn't perfect. The problem was that Copernicus was still assuming perfectly circular orbits for all of the planets. It would be another seventy years before Johannes Kepler would fix this problem by introducing the idea of elliptical orbits.

To say that Copernicus waited until the last possible moment to publish his theories is a bit of an understatement. In addition to publishing *On the Revolution of the Heavenly Spheres* in 1543, he also died in 1543. One biography of Copernicus claims that he received the final printer's proof sheet on his death bed.[18]

When it was finally published, Copernicus's work would become one of the most influential books in the history of astronomy. But it was also a masterclass in ecclesiastical bet-hedging. The printed volume included a copy of Cardinal Schönberg's letter and an additional introduction by the Lutheran theologian Andreas Osiander explaining that this was a hypothetical model and not necessarily true. Osiander oversaw the actual printing of *On the Revolution of the Heavenly Spheres*, so it is unclear if Copernicus was aware that Osiander had added himself into the finished book.[19] For good measure, the book was dedicated to Pope Paul III, the successor to the pope who had allegedly been so happy to learn about Copernicus.

In his dedication to the pope, Copernicus lays out exactly why he undertook writing *On the Revolution of the Heavenly Spheres*. He explains that it is not meant as an attack on the Church. To that end, he specifically mentions the patronage of Cardinal Schönberg as well as the support he received from the bishop of Chelmno. He frames his work as purely astronomy and asks that it be judged that way. Finally, he points out how useful good astronomical models can be to the church as they pursue calendric reforms.[20]

In short, Copernicus went to great pains to temper the papal response to his model. But none came. Maybe Pope Paul III shared Clement VII's enthusiasm for the new cosmology. Maybe he just had bigger fish to fry. In 1543, Protestantism was spreading through Europe and King Henry VIII of England had just finished up his second divorce. Whatever the reason, Copernicus's ideas were met with relative papal silence. But then in 1616, seventy-three years after its publication and eighty-two years after its premise had delighted Pope Clement VII, *On the Revolution of the Heavenly Spheres* was added to the Catholic Church's *Index of Forbidden Books*. But that had far less to do with Copernicus and more to do with Galileo.

Galileo's Problems Begin

Galileo Galilei was born in 1564, two full decades after Copernicus published *On the Revolution of the Heavenly Spheres.* We know that early in his career, Galileo believed the earth was in the center of the universe. We know this not from his studies of the heavens, but from his studies of hell. When he was in his twenties, Galileo was an art instructor teaching painting techniques at the Florentine Academy, with a side interest in geometry and mathematics. One of his earliest surviving works is a pair of lectures that he gave there in 1588 on the size and shape of the inferno as described in Dante's *Divine Comedy.* Curiously, his geometric analysis begins with "let us imagine a straight line which comes from the center of the earth (which is also the center of heaviness and of the Universe) to Jerusalem."[21] Galileo started from the assumption that the earth is at the center of the universe. But that does not necessarily mean that he was working from the Ptolemaic model. By 1588, there was a new model of the universe, competing with both Ptolemy and Copernicus, and trying to combine the best aspects of each.

Ptolemy and Copernicus both had pragmatic truth on their side. You could use either model to determine (roughly) where in the sky things would be on any given night. Ptolemy had coherent truth in that his model agreed with other things that astronomers believed about the universe: that the earth was firmly planted in the center of the cosmos. This idea came from Aristotle and was also strongly implied by the Bible. The Copernican model also had coherent truth, but it cohered with a different set of beliefs that astronomers held. Because it could explain the retrograde motion of the planets without relying on epicycles and other geometric constructs invoked by Ptolemy, it cohered with the idea that all heavenly motion was simple and circular. In the 1570s, Danish astronomer Tycho Brahe created a model of the universe that combined the philosophical coherence of Ptolemy with the mathematical coherence of Copernicus.

The Tychonic Model (sometimes called by the clunky name *geoheliocentrism*) keeps an unmoving earth at the center of the universe. (See figure 3.3.) Brahe admired the mathematical simplicity of Copernicus but believed that a stationary earth was just common sense based on our experience.[22] The sun, moon, and stars orbit the earth, but then the other planets of the solar system orbit the sun. It's an ingenious model that solves the problems of both Ptolemy and Copernicus, and it enjoyed a brief popularity in the late sixteenth century.

Whether Galileo believed in a Ptolemaic or a Tychonic universe when teaching at the Florentine Academy, we know that by the 1590s, he was firmly on board the Copernican bandwagon. And we know that Galileo came to accept heliocentricity because it did a better job of explaining his observations. In a 1597 letter to the German astronomer Johannes Kepler, Galileo writes, "I

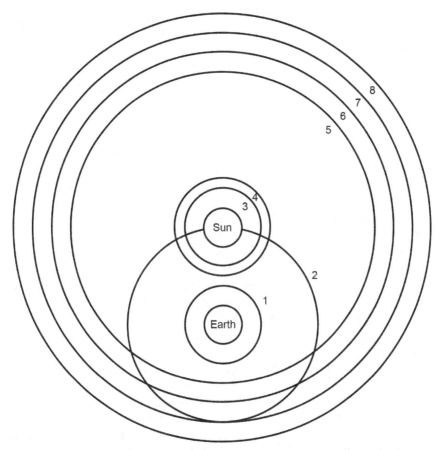

Figure 3.3.   The Tychonic model of the universe. In this model, the unmoving earth is orbited by (1) the moon and (2) the sun. The sun is in turn orbited by (3) Mercury, (4) Venus, (5) Mars, (6) Jupiter, (7) Saturn, and (8) the stars. This model eliminates the need for epicycles while still keeping the earth still and (mostly) central.

accepted the Copernican position several years ago and discovered from thence the causes of many natural effects which are doubtless inexplicable by the current theories."[23] He continues to tell Kepler that, although he has written up a manuscript detailing his discoveries, he is concerned about the ramifications of publishing. Copernicus's theories may not have faced a backlash from the Catholic Church, but his reputation as a mathematician was diminished in some circles by their publication. Galileo wanted to avoid this fate if possible.[24]

Galileo did eventually begin to share his astronomical observations. In 1610, he published *Sidereus Nuncius* (usually translated into English as either

The Sidereal Messenger or *The Starry Messenger*), detailing observations that he made through the telescope. Contrary to popular belief, Galileo did not invent the telescope, but he did make some significant improvements to it. Over a six-month period in 1609, he improved the magnification powers of his telescope from the then-common 3x power to 20x power.[25] In *The Starry Messenger*, Galileo describes being able to see mountains on the moon, clouds of distant stars, and four moons orbiting around the planet Jupiter.[26]

Galileo's discovery of the moons of Jupiter is often cited as evidence for the Copernican model, but that's not necessarily true. Obviously, in a Ptolemaic universe, the discovery of moons orbiting Jupiter is a big problem. After all, according to Ptolemy, everything revolves around the earth. But in the Tychonic system, Jupiter having moons is not necessarily problematic. The sun orbits the earth, and Jupiter orbits the sun. So why can't something else orbit around Jupiter? As a matter of fact, it was not anything at all that Galileo saw through his telescope that convinced him that the earth moved around the sun. He drew that conclusion from observations right here on earth.

Galileo was one of the first people to suggest that the tides are caused by the motion of the earth.[27] And the Copernican model of the universe was the only one at the time in which the earth moved. For both Ptolemy and Tycho Brahe, the earth was not only central to the universe but also unmoving. The belief that the earth neither moves nor rotates is called *geostaticism*. It was an appealing belief in Galileo's day, in part because the earth doesn't feel like it's moving. Geostaticism was also a common belief because it is consistent with several different passages in the Bible that clearly say the earth does not move. As an example, 1 Chronicles 16:30 says, "The world is firmly established; it shall never be moved." The discoveries that Galileo made with his telescope surprised people, but they were difficult to deny when anyone could look through the telescope and see them for themselves. Galileo's assertion that the earth was in motion made people more uncomfortable. Again, this claim seemed to defy scriptural authority outright. Galileo, however, had a rather elegant explanation for why the earth's motion shouldn't be as concerning as people were making it.

In a 1613 letter to his friend Benedetto Castelli, a monk and mathematician, Galileo dismissed claims that his advocacy of heliocentricity was heretical. He pointed out that Copernicus was himself a priest (although on this point the historical record is ambiguous) and a canon of the Church, and that *On the Revolution of the Heavenly Spheres* was published with the support of the Church. He also said that where there was a conflict between our God-given senses and ambiguous passages in the Bible, we should interpret those passages to align with our observations. Finally, Galileo wrote that the Bible's primary purpose is to help humanity reach salvation and to tell us what we must believe to be saved. Based on this, he argued, if a particular belief about the structure of the solar system was necessary to our salvation, then the structure of the solar system would be explicitly detailed in the Bible. These arguments won over Castelli, and would later form the basis for a longer work laying out a similar argument to his patrons.[28] But not everyone was so convinced.

Galileo's troubles began when he caught the attention of Father Tommaso Caccini. Caccini was an ambitious Dominican friar who imagined himself going down in history as a great orator and theologian. Even his name reflected his ambition. He was born Cosimo Caccini in 1574 but took the name Tommaso upon his ordination as a sign that he hoped to follow in the illustrious footsteps of St. Thomas Aquinas.[29] Galileo had been caught in priestly crosshairs before Caccini, but those instances had mostly come to nothing.

Caccini began his attacks on Galileo during an advent service in December of 1614. The sermon itself is lost to history, but it reportedly began with the story of Joshua stopping the sun in the sky, moved on to an attack against any science or even math that contradicted the literal word of the Bible, called out Galileo by name as a heretic, and ended with a fiery cry of a passage from the Acts of the Apostles: "Men of Galilee, why do you stand looking up toward heaven?"[30] Galileo had supporters in the Church, but Caccini had passion and connections and was eventually able to bring his arguments before the pope and the Inquisition.

In 1616, less than a century after Copernicus dedicated his theories to Pope Paul III, Pope Paul V declared them to be heretical. As a result, Paul V is often portrayed as one of the great villains in the history of science, but his relationship with Galileo was actually a bit more complicated than that. Before Paul V became Galileo's antagonist, he was Galileo's sponsor. One of the first great scientific societies in European history, the Lincean Academy (or the Academy of the Lynx-Eyed), was organized in the Papal States in 1603. The Linceans believed first and foremost in the importance of observation as a mode of scientific inquiry and modeled themselves after the lynx, a creature that medieval bestiaries proclaimed to have the greatest eyesight in the animal kingdom. Based on the observations that Galileo had published in *The Starry Messenger*, he was an obvious candidate for membership in the Lincean Academy, and in 1611, Pope Paul V sponsored him for membership.

Pope Paul V respected Galileo's powers of observation but could not ignore Galileo's arguments. In 1615, a group of Dominican friars brought the pope a copy of Galileo's letter to Castelli. Since the letter called for people to interpret ambiguous Bible passages in ways that were consistent with scientific observation, it presented a serious departure from Catholic doctrine. In the previous century, in an effort to stem the rising tide of Protestant reform in Europe, the Council of Trent had made it heresy for anyone outside of the Church hierarchy to interpret scripture.[31]

When Caccini heard that Galileo's letter had found its way into the pope's hands, he went to Rome. He testified before the Inquisition that if Copernicus's theory required people to interpret the scriptures for themselves, then it must be considered heretical. This is a classic example of a logical fallacy called the argument from authority. In this fallacy, you judge an argument based on the authority of the person making it, rather than on its own merits.[32] What Galileo and Copernicus were arguing didn't actually matter in this case. If they were on one side, and the Catholic hierarchy was on the other, then they were wrong. In 1616, the Inquisition ruled in agreement with Caccini and added *On the Revolution of the Heavenly Spheres* to the *Index of Forbidden Books*. Pope Paul V sent a message to Galileo informing him that the opinion that the earth goes around the sun was heretical and that he could not defend it, either in writing or orally, without risking severe punishment.[33]

The Pros and Cons of Powerful Friends

Maffeo Barberini was born in Florence in 1568. At an early age, he was sent to Rome to live with his uncle, a papal official. Thanks to his uncle's power and influence, he had the best education money could buy and spent time in the royal courts of Europe. By the time he was 40, he had inherited his uncle's fortune and had been made a cardinal by Pope Paul V. It was in his capacity as cardinal that Barberini first met Galileo.

At a dinner held in Cardinal Barberini's palace, Galileo got into a heated debate with other guests about why things float.[34] This was in 1611, after the publication of *The Starry Messenger*, but before the banning of Copernicus. At the time, Galileo was a prominent scientist and a newly minted member of the Lincean Academy visiting Rome and enjoying his celebrity. Cardinal Barberini was impressed by Galileo's argument as well as his aggressive debating style and instantly became a great admirer of his distinguished guest.

In 1616, when Galileo was censured by the Inquisition, Cardinal Barberini was a dissenting voice and made known his general disdain for the outcome.[35] In 1620, at a time when Galileo was considered anathema to most of the Catholic hierarchy, Cardinal Barberini praised him in poetry.[36] And so, it seemed an amazing opportunity for Galileo to get back in the good graces of the Catholic Church when, in 1623, his longtime friend and benefactor Cardinal Maffeo Barberini was elected Pope Urban VIII.

Having a friend and supporter as pope significantly changed Galileo's prospects. During the first years of Urban VIII's papacy, Galileo had at least six papal audiences in which he felt free to discuss his scientific works. The pope had a long-standing respect for Galileo as a scientist and knew that he had wanted for some time to write about the Copernican model of the universe. However, he also knew that he was bound by the interdiction against advocating heliocentricity put into place under Paul V.

Eventually, Pope Urban and Galileo reached an agreement. Galileo was barred by the Inquisition from believing in heliocentricity and from advocating for it, but not from simply describing it. Pope Urban told Galileo that he would give his permission for the publication of a book if Galileo could write it while abiding by these rules. That is, Galileo could describe the Copernican system, but he couldn't say that it was the correct model of the universe. Any evidence he gave in favor of heliocentricity must be balanced by equal evidence for geocentricity. Galileo gladly accepted this opportunity.

Galileo's deal with the pope resulted in the 1632 publication of *Dialogue Concerning the Two Chief World Systems* complete with a stamp of official imprimatur on its title page.[37] Along with its scientific importance, *The Dialogue* (as it has come to be known) helped to create the archetype for a form of popular scientific presentation that is still with us today—the artificial debate. News sources, in an attempt to appear free of bias or advocacy, will create a false sense of balance by bringing on guests to promote "both sides" of an issue regardless of whether or not there is equal validity to both sides.[38] This is a form of an older logical fallacy called the argument to moderation, which goes beyond just creating false debates by assuming that when people are arguing two wildly divergent points of view, the truth must be somewhere in the middle. One might consider the Tychonic model of the solar system, with its combination of things moving around the earth and the sun, to be an astronomical example of the argument to moderation.

In order to comply with Pope Urban's restrictions, Galileo wrote *The Dialogue* in the form of a debate between two different astronomers each trying to convince a third party of the correctness of their model of the universe. Ostensibly, this is a clever way to ensure that each side is equally represented. Unfortunately, Galileo couldn't help but advocate for heliocentricity in some very obvious ways.

The first hint that the reader gets that things will not go well for the Ptolemaic model of the universe comes in the names of the characters interacting in *The Dialogue*. The debate takes place in the palace of a nobleman named Sagredo, who also plays the part of the neutral third party looking to be convinced. Sagredo is named for Galileo's personal friend, the mathematician Giovanni Francesco Sagredo.[39] Sagredo's first guest in *The Dialogue* is the Copernican proponent Salviati. Salviati was named for Filippo Salviati, another friend of Galileo's and a fellow member of the Lincean Academy.[40] However, it is not a coincidence that Salviati's name also comes from an Italian word meaning "sage." Where Galileo really crossed the line was in naming the character who would be defending the Ptolemaic universe. That poor character is named

Simplicio. Galileo claimed that Simplicio is named for Simplicius of Cilicia, a fifth-century philosopher who wrote extensive commentaries on Aristotle's astronomical work,[41] but the name also means "simple" in Italian.

Beyond just framing *The Dialogue* as a debate between "the sage" and "the simple," Galileo doesn't make it anything close to a fair fight. Salviati wipes the palazzo's floor with Simplicio. *The Dialogue* is divided into four separate days of discussion, and by day 4, Simplicio is mostly silent while Salviati is given long monologues about the tides. To add insult to injury, Salviati (like Galileo) clearly considers the tides to be the single most compelling piece of evidence for the earth's motion. But it is clear that Simplicio doesn't understand the tides at all. He is an hour late to the discussion on day 3 because his gondola was stuck in an ebb tide.[42]

There are several myths and legends surrounding the fallout from Galileo's publication of *The Dialogue*. Some sources claim that Pope Urban had Galileo taken to a torture chamber and merely shown the machines because as a man of science Galileo would be able to figure out for himself how they work. Other sources have Galileo recanting his heresy before the Inquisition but then defiantly muttering "*eppur se muove*" (but it does move) on his way out of their presence.[43] There is probably little truth to either of these stories. What is definitely true, however, is that *The Dialogue* broke the friendship between Galileo and Pope Urban.

In 1633, Galileo was found guilty by the Inquisition of violating their order of 1616 by advocating for heliocentricity, and of holding the heretical beliefs that the earth moves but the sun does not. His sentence was threefold. He was to formally renounce his heretical beliefs. He was to be imprisoned indefinitely. And no more books of his—past, present, or future—would receive permission to be printed ever again.[44] Galileo's imprisonment was commuted to house arrest, but the rest held. He signed a formal abjuration and withdrew from public life. The only other work of his that was ever published again in his lifetime was a book on mechanics that was smuggled out of his house by a friend and published in Holland.

What might be the most tragic aspect of Galileo's second inquisition is how unnecessary it was. It was entirely possible to present Copernicus without advocating for it. Catholic universities throughout Europe used it as a mathematical model to teach geometry as early as the 1600s.[45] Furthermore, when we praise Galileo in hindsight, it is easy to lose track of a few crucial points. The theory he was advocating did not hold a scientific consensus at the time. The Tychonic model was a popular alternative to Copernicus during Galileo's lifetime. Also, while the Copernican model did have some advantages over Ptolemy from the get-go, there was one crucial question about the stars and planets that it couldn't answer any better. Why would anything in the heavens be moving at all?

Isaac Newton Explains It All

The most traveled explorer in human history wasn't human at all. On September 5, 1977, the space probe Voyager 1 left Florida and never came home. Armed with a radio, a few cameras, and some plutonium-powered batteries, Voyager 1 has now traveled over 14 billion miles and is more than 150 times further from the sun than the earth is. On Valentine's Day, 1990, Voyager 1 took one last look back toward home before leaving our solar system forever. The picture it took on that day is called "The Pale Blue Dot," because from that distance that's all the earth appears to be—a tiny speck barely a pixel in size. Neither the moon nor the sun is in the field of view of the picture. That picture was the first, and so far the only, picture that has ever been taken of our solar system from the outside.

Diagrams and models of our solar system are so ubiquitous that it is easy to forget that they are just diagrams and models. The Apollo astronauts were able to stand on the moon, look up at the earth, and say "Yep. It's round." But nobody has ever stood back sufficiently far enough to watch the earth go around the sun. So why can we be so sure today that it does? Once again, it comes down to coherence truth.

In 1687, Isaac Newton made the universe a simpler place. The "Rules for Reasoning" that he detailed in his *Philosophiæ Naturalis Principia Mathematica* (more commonly called *The Principia*) encompassed one of the first formal claims that all objects have universal properties and are subject to the same forces. In retrospect, some of these rules seem painfully obvious (e.g., if you keep seeing the same effect, assume a constant cause),[46] but at the time, they were revolutionary. They enabled scientists to take the lessons they learned studying one system and apply those lessons to other systems.

This shift in scientific reasoning was particularly useful for astronomers. Rule 3 says that if you are unable to measure a property in some body, but that property is constant in everything you can measure, then you may assume that property to be universal in all bodies.[47] In other words, if everything on earth has mass, then everything everywhere has mass. If everything on earth is affected by gravity, then everything everywhere is affected by gravity.

Prior to Newton, every model of the solar system glossed over one important question: Whether you believe the sun is moving or the earth is moving, why is it moving? Ptolemy had a beautiful system of circles moving within circles, but no particularly compelling explanation as to why things moved in a circular path, let alone why they would move in a circle-within-a-circle path. Tycho Brahe had some things orbiting the earth and others orbiting the sun but no explanation of the determining factor in which body orbited which other body. Copernicus had the simplest model, with a stationary sun and everything else moving around it, but why would the sun be still when nothing else was?

There is some dispute as to exactly how much credit Newton deserves for his various discoveries. Some people want to give the German mathematician Gottlieb Leibniz priority for the development of calculus. Robert Hooke proposed a law of constant motion and an inverse-squares gravitational law twenty years before Newton published *The Principia*. But by universalizing the laws of nature, Newton gave us a guideline for definitively choosing between cosmological models. The motion of heavenly bodies is just like the motion of earthly bodies. Therefore, whichever model of the solar system can best be explained using the earthly laws of gravity should take priority. And on that criterion, Copernicus stands above the rest. But this ultimate evidence for heliocentricity was not available to Galileo.

Newton changed the question from "Does the earth go around the sun?" to "*Why* does the earth go around the sun?" and provided us with an easy answer. The earth is a massive object caught in the gravitational pull of a more massive object. The consequences of this arrangement, heliocentricity, are inevitable, and no more controversial or mathematically complicated than trying to figure out where a cannonball would land.

Forgiving Galileo but Not Always Believing Him

Not all religious people think the same way about science. Not all Christians think the same way about science. Not all Catholics think the same way about science. Not all popes think the same way about science. And as we have seen, not even all popes named Paul have thought the same way about science. In a post-Newtonian world, as it has become increasingly clear that the earth does, in fact, move around the sun, multiple new popes have incrementally softened the Catholic Church's official stance on Copernicus and Galileo.

In the 1710s, Pope Clement XI lifted the Inquisition's ban on publishing the works of Galileo.[48] By the 1830s, Copernicus and other works on heliocentricity were removed from the *Index of Forbidden Books*.[49] In the 1890s, Pope Leo XIII reestablished the Vatican Observatory as part of his efforts to show that science and religion could not only coexist but also thrive together.[50] And in 1992, Pope John Paul II described the Galileo Affair (as it has come to be called) as a mistake and, astonishingly, quoted from Galileo's letter to Castelli, describing it as "a small lesson in biblical hermeneutics."[51] The same position that had initially resulted in Galileo's inquisition had gained the endorsement of the papacy itself.

Today there is very little, if any, mainstream religious objection to heliocentricity. And the structure of the solar system is definitely being taught in schools. It appears repeatedly in the Next Generation Science Standards that have been adopted by dozens of states[52] and is featured in all of the most commonly used science textbooks. So why are 28 percent of Americans unsure about it?[53] It's not just a fluke or a bad test day. The results of that survey are repeatable over time. A similar study from 2014 shows almost the exact same results. In that, 26 percent of Americans got that same question wrong.[54]

One possible reason that so many Americans got the question wrong could be the American education system. After all, our school system is a favorite target of scorn whenever we seem to be lagging behind the rest of the world in any kind of knowledge. And sure enough, when the 2014 study was first published, several news outlets lamented the results, and many of those stories included some sort of reference to education. But this argument doesn't hold up to scrutiny.

The 2014 survey included information on how well citizens of other countries performed on the same survey. Only 74 percent of Americans knowing that the earth goes around the sun may seem dismal, but it was a high enough mark that we outperformed citizens of Malaysia (72 percent), India (70 percent), and the EU (66 percent).[55] It's not just Americans who are confused about the structure of the solar system. It's people in general. And the problem, most likely, comes down to our innate desire to believe our own senses.

In 1620, the English philosopher Francis Bacon published his masterwork, *Novum Organum* (or *The New Toolkit*). In it, he laid out what he felt were the most important mental tools that people had for sensibly understanding and explaining the world. However, he also made special note of the peculiar biases and hang-ups that people have that keep them from understanding. He called these the "Idols of the Mind," and two of them come into play here.[56]

The Idols of the Tribe are the ways that we misprocess information just because we are human and bring a human perspective to our observations and classifications.[57] For example, consider the following statement: "Squirrels are incredibly large animals." At first glance, it seems ridiculous, but it is undeniably true. One-quarter of all named species are beetles. Not insects, beetles. The world's ants outweigh the world's humans by a factor of ten.[58] The oceans around Antarctica are home to an estimated krill population of half a quadrillion.[59] That's 500,000,000,000,000 krill. With those numbers for context, it's obvious that squirrels are easily within the top 1 percent of animals that have ever lived on earth size-wise. But they are smaller than us, so we have trouble seeing them for the giants they are.

The Idols of the Cave (sometimes translated as Idols of the Den) are the personal preferences that each individual brings with them when interpreting the world.[60] We like explanations that we can understand, either because they are simple, because they appeal to our interests, or sometimes because they are flattering. In the 1980s, when geologists first started to find compelling evidence that an asteroid impact killed the dinosaurs (more on that in chapter 6), the scientific community was divided over how readily they accepted that model. Astronomers, generally speaking, thought it was a compelling explanation. Paleontologists were less thrilled, since it seemed to move the event out of their wheelhouse. Volcanologists were downright resistant to the idea. Many had built careers explaining how volcanoes had killed the dinosaurs. But the general public received the theory very well. It even made the cover of *Time* magazine. It was a simple explanation that everyone could understand. And it was cool!

Having a stationary earth in the center of the universe is contrary to our best mathematical models and to our understanding of gravity, but it appeals to both the Idols of the Tribe and the Idols of the Cave. It confirms what we experience and observe, in that the earth does not feel like it is careening through space. It means that the sunrise is a glorious harbinger of a new day, not an optical illusion caused by relative motion. It is simple. And above all, it is flattering. The whole of reality literally revolves around us.

CHAPTER 4

The Earth Is Solid

It's uncomfortable inside the earth. It's hot and it's dark, and things are only more uncomfortable the deeper down you go. At the bottom of South Africa's Mponeng gold mine (the deepest mine anywhere on earth), the walls of the mine shaft reach a blistering 60° C (140° F).[1] The Mponeng mine is one of the deepest places that humans have directly explored, but it is still less than three miles deep. For comparison, a Jules Verne–style journey to the center of the earth would be nearly four thousand miles. Despite the fact that our direct explorations of the earth's depths have been quite limited, our knowledge of what's down there is considerable. Developing that knowledge was a matter of finding the right tools to allow us to explore the earth's depths from the comfort of its surface.

Edmond Halley and the Wandering Pole

Astronomers in the 1600s had a serious problem. North seemed to be moving. Not true north, as defined by the rotation of the earth, but compass north. Even back then, people had already known for at least one hundred years that they were two different things. They just assumed that there was something large and magnetic near the north pole that compasses conveniently pointed to. At least one map at the time even included an *Insula Magnetūm* (Island of Magnets) located near the geographic north pole.[2]

Whatever it was that was drawing compass needles north-ish, by the late 1600s, it was very obviously moving. In 1692, Edmond Halley (best remembered today for his eponymous comet) wrote a paper cataloging how compass readings had changed over the previous 150 years.[3] He also proposed a remarkable explanation for the phenomenon. According to Halley, we live on the outermost of a nested set of hollow earths.

Halley asserted that nothing on the surface of the earth was actually moving at all. He also claimed that if it were the earth itself that was magnetic, then the pole of that magnet shouldn't be moving either. Instead, Halley said that there is another earth inside of ours. Our earth is hollow, with an outer shell about five hundred miles thick. On the inside of that shell is another atmosphere and within that another earth. If that other earth is rotating similarly to our earth, but not exactly identically, then its north pole would move relative to our north pole, which would confuse compass needles.

Halley didn't actually stop there. In all, he claimed that our planet is a nested set of four earths. The inner worlds, he assumed, corresponded in size to the planets Venus, Mars, and Mercury. From here, Halley put his powers of deduction to use. He reasoned that since nothing in this universe is made without a purpose, these inner earths must serve the same purpose that the outer earth does—to harbor life. Since you can't have life without sunlight, he reasoned that the internal surface of each earth must glow like the surface of the sun.[4]

In 1716, nearly a quarter-century after he initially proposed his model of multiple, internally luminescent, hollow earths, Halley was able to add to his theory and publish what he saw as proof of its reality. In March of that year, there was a particularly strong aurora borealis. It was visible as far south as London, allowing people who had never seen the aurora before to see it and wonder about its origins. Halley proposed that the lights that people were seeing in the sky were the earth's internal light seeping out from below. Based on the angle that the lights made with the surface of the earth, it was obvious to Halley that some fissure had opened in our earth near the poles, temporarily allowing the inner light to emanate out.[5]

Halley's model of a hollow earth may seem strange, but it is enduring. According to the Library of Congress, no fewer than nine different books proposing a hollow earth theory, each by a different author with different theories and new evidence, were published just in the twentieth century.[6] Interestingly, many of these books included Halley's idea that there was an opening to the interior near the poles. There was, therefore, a particular abundance of these books being published at about the time that polar explorers Peary and Amundsen were making their ways to the north and south poles. Some books accused the polar explorers of covering up their discoveries of the inner earth. Others claimed the expeditions themselves were complete frauds.

Moby Dick, Santa Claus, and John Quincy Adams Meet the Hollow Earth Theory

Perhaps the most influential hollow-earther (other than Halley) was John Cleves Symmes Jr. (1780–1829). Symmes was a hero in the war of 1812, and later a merchant in St. Louis, so it's not entirely clear where his interest in astronomy and geology came from. But, in 1818, Symmes published his "No. 1 Circular." It was ambitious to say the least. It began, "I declare the earth is hollow, and habitable within; containing a number of solid concentrick spheres, one within the other, and that it is open at the poles 12 or 16 degrees; I pledge my life in support of this truth, and am ready to explore the hollow, if the world will support and aid me in the undertaking."[7]

The model that Symmes was proposing is that the earth is less of a sphere and more of a donut. Not only is there no magnetic island at the north pole, but also there is in fact nothing there but a gaping hole. Symmes also proposed a similar opening at the south pole. Like Halley, Symmes thought these openings allowed an inner heat to radiate out so that the area around the poles was actually quite temperate and comfortable.

Symmes had an advantage when it came to getting his theories tested. And it was not his scientific credentials or his passionate rhetoric. It was his family. His uncle (for whom he was named), John Cleves Symmes, was a colonel in the revolutionary war, a justice on the New Jersey supreme court, and a delegate to the continental congress. His cousin Anna Symmes would eventually become first lady of the United States, albeit briefly. She was married to William Henry Harrison, who died before the happy couple ever set foot in the White House.

Thanks to his family name, as well as his own background, Symmes had the political connections to get things done. On February 7, 1823, Senator Benjamin Ruggles of Ohio introduced a motion on behalf of his constituents to put Symmes in charge of an expedition to the poles. Symmes also gained the support of at least one newspaper editor, J. N. Reynolds of the Wilmington, Ohio, *Spectator*. Reynolds is something of a mysterious character, down to the fact that nobody is entirely sure what the *J. N.* stood for. Most sources report that the *J* is for Jeremiah, but at least one obituary claimed it was James. We know he attended Ohio University but are unsure if he graduated. What we do know is that in 1824, Reynolds left his newspaper to begin a speaking tour with Symmes.[8]

Symmes and Reynolds lobbied tirelessly for America to explore the South Pacific and, after that, the south pole. Eventually, their plan found favor with President John Quincy Adams. At the behest of the secretary of the Navy, Reynolds drew up a detailed plan to explore the southern oceans.[9] In his 1828 State of the Union address, Adams declared, "The vessel is ready to go."[10] However, after John Quincy Adams lost the 1828 election, his successor, Andrew Jackson, had no interest in nautical exploration, and the project never actually happened.

Symmes died in 1829, having never made it any closer to the north pole than southern Ontario. While Symmes's scientific legacy is minimal, his influence in popular culture may be far more significant. Symmes promoted his theory of a hollow earth with temperate poles in the mid-1820s, at about the same time Clement Moore first published his classic poem "A Visit from St. Nicholas." The coincidence has caused at least one author to suggest that Symmes's model inspired the belief that Santa Claus can live and work comfortably at the north pole.[11]

The circumstances of J. N. Reynolds's literary legacy may be even more extraordinary. He eventually became disenchanted with the idea of a hollow earth, but his passion for exploration persisted. After President Jackson halted the American voyage to the South Pacific, Reynolds raised the money for a private voyage instead. It departed New York in 1829 and caught sight of the Antarctic coast later that year. Unfortunately, the trip was so difficult that, on the return voyage, the crew mutinied and put Reynolds ashore in Chile. Reynolds remained in Chile for two years before being picked up by the U.S. naval ship *Potomac* in 1832. During the two years Reynolds spent on the *Potomac*, as it circumnavigated the globe, he gathered sailors' stories of a terrible white sperm whale that lived off the coast of Chile and couldn't be killed. Upon finally returning home, Reynolds published those stories in *Knickerbocker* magazine under the title *Mocha Dick: or The White Whale of the Pacific*,[12] which would eventually inspire a similarly named novel by Herman Melville.

Turning the Hollow Earth Theory Inside Out

If the earth is round, and it careens through space while simultaneously spinning like a top, why don't we fall off? It seems like an obvious question, and in 1869, physician and self-proclaimed alchemist Cyrus Teed proposed an incredibly novel, and remarkably wrong, answer. Teed lived in Utica, New York, where he practiced "eclectic medicine," which meant he was just as likely to treat patients with herbs or electric shocks as actual medicine. One afternoon, after accidentally electrocuting himself into unconsciousness with one of his instruments, he had a vision of the earth's true nature. According to Teed, we don't fall off the earth because we don't live *on* the earth. We live *in* the earth.[13]

Like Halley and Symmes, Teed believed that the earth is hollow. But uniquely, Teed added the wrinkle that the surface we live on is not the convex outside of the sphere. We live on the concave inside of the sphere. At the center of that sphere is the sun. As the earth spins on its axis, it is centrifugal force, not gravity, that holds us to the ground. Teed's model, sometimes called "Sky-centrism," has a lot of logical problems. One obvious problem is the existence of night. A concave earth would have no horizon to block our view of the sun. Just like different constellations in the southern and northern hemispheres are a challenge for a flat earth model, if we lived inside a sphere with a sun at its center, that sun should always be visible. Teed tried to get around this detail with an elaborate model of the inner sky that included mirrors and invisible batteries, but nothing he came up with could ever credibly explain the darkness of night.

Despite the seemingly undeniable existence of night, Teed gained quite a following. Upon recovering from his electrically inspired revelation, Teed changed his name to Koresh and founded a Utopian community whose population numbered in the hundreds and practiced Koreshanity—a faith that combined alchemy, skycentrism, transmigration of souls, and of course, a belief in the immortal divinity of Teed himself.[14] In the 1890s, Teed and the entire Koreshan Unity commune moved to Florida, where they were given some land by a local rancher and established the town of Estero.

Like Samuel Rowbotham before him, Teed may have chosen the site for his utopian community for scientific purposes. Florida is remarkably flat. Its highest natural point is only about three hundred feet above sea level. This is because Florida is very different geologically from the rest of the United States. Unlike most of North America, which is formed from typical continental granite, Florida is made of limestone. Limestone only forms in warm ocean water and forms best when the water is shallow. When limestone is exposed above sea level, it erodes fairly quickly. As a result, the elevation of a limestone platform like Florida stays relatively flat, and very close to current sea level. With so much flat countryside to work with, Teed had the opportunity to perform experiments similar to those at the Bedford Canal. But Teed was looking for something very different than Rowbotham was. Whereas Rowbotham was expecting the earth to fall away from the horizon, Teed was expecting the earth to curl upward the farther away you looked.

Perhaps inevitably, by establishing Estero as a home for skycentrists, Teed drew the ire of the flat-earthers in the Zetetic Society. The feud became all the more bitter when Ulysses Morrow, one of the leaders of the Zetetic movement, defected to Estero. Before moving to Estero to work with Teed, Morrow had spent several years in Pennsylvania proselytizing for the Zetetic Society and Rowbotham's model of the universe. What caused his change of heart regarding the earth's shape is unclear, but it certainly seems to have caused quite a crisis for the Zetetics. The first 1897 issue of *Earth (Not a Globe) Review* is a detailed rebuttal of Koreshanity in general and a refutation of Morrow in particular.[15]

In an effort to prove his model of the universe, Teed commissioned Morrow to lead the Koreshan Geodetic Survey—an organized attempt to empirically measure the concavity of the earth. In 1897, Morrow built "the Rectilineator"—a series of twelve-foot brass and mahogany devices designed to create straight lines of sight over great distances. Using the Rectilineator and a small army of volunteers, Morrow claimed to successfully measure the internal convexity of the earth over a four-mile stretch of beach in Naples, Florida. Conveniently, his notes were archived and never published.

Continuing to claim that we live inside the earth without presenting the alleged evidence from the Rectilineator is a variation on the argument from ignorance.[16] In this case, the Koreshans were asserting that something was true simply because it had not been proven false, while simultaneously claiming that they had the evidence but weren't sharing it. Under these circumstances, it is also fair to subject the Koreshans to a doctrine created by the author Christopher Hitchens and sometimes referred to as Hitchens's Razor: "What can be asserted without evidence can also be dismissed without evidence."[17]

On December 22, 1908, in direct defiance of Koreshan doctrine, Cyrus Teed died. Over time, the Koreshans eventually drifted away from Estero, with the last one finally leaving in the 1960s. The land they used to own is now a historic site and state park.[18]

When Edmond Halley set out to solve the mystery of the migrating pole, he inadvertently set off a two-hundred-year hunt for a polar entrance to the earth's interior. He did, however, get one very important thing correct. Determining why the magnetic pole moves would require an understanding of the earth's interior. Unfortunately, it would take geologists nearly three hundred years to understand the relationship.

Weighing the Earth to Find Out What's Inside It

As a species, we have done a pretty good job of exploring the earth's surface. But the earth isn't a surface. It's a sphere. And the interior of that sphere is largely unexplored. The deepest hole that exists anywhere on earth is in Russia. The Kola Superdeep Borehole was begun in 1970 as an experiment simply to see what was down there. By the time it was abandoned due to lack of funding in 1995, it had reached a depth of 12,262 meters (about 7.5 miles).[19] According to our best geodetic models, the average radius of the earth is 6,378.1 km (a little less than four thousand miles), meaning that in twenty-five years of trying, we drilled less than a quarter of a percent of the way to the earth's center. Yet despite our lack of direct observation, we actually know quite a bit about the earth's interior.

As a general rule, you don't need to open a box to determine whether or not it is empty. If the box is heavy, there is probably something in it. The same principle holds true for determining whether or not the earth is hollow. If you could find a way to measure the weight of the earth, that would give you a pretty good idea of what must be on the inside. Weighing the earth isn't as easy as just putting it on a scale, but it is possible.

Any measurement requires units. A person doesn't just weigh 180. They weigh 180 pounds. In practical terms, what that means is that if you had a supply of one-pound weights, you would need 180 of those weights to make a pile equal in weight to the person. So when we say that a person weighs 180 pounds, we are saying that the ratio between the weight of the person and the weight of a pound is 180. You could use other units of measure, and that would produce other ratios. That same person would weigh 12.9 stone or 2,880 ounces, depending on your unit of choice. In the 1770s, a team of scientists and surveyors from the Royal Academy tried to measure the mass of the earth not in pounds or kilograms, but in mountains.

Measuring the mass of a mountain is hard. Measuring the mass of the earth is even harder. Measuring the ratio of their masses is much easier, but the method for doing it sounds almost absurd. Imagine a weight tied to the end of a string. That string will hang straight down under the influence of gravity. But if there were a nearby massive object, like a mountain, that mountain might have enough mass to produce its own gravity. The gravity of the mountain would therefore also be pulling on the weight, and the string would be deflected just slightly away from straight down in the direction of the mountain. In other words, if you are close to a large enough mass, like a mountain, things don't fall straight down. They fall down and a little bit toward the mountain. The larger the mass of that mountain is, the more that falling object's path will deviate from straight down.

The idea that large terrestrial objects could create gravitational fields strong enough to compete with the earth's own gravity was first suggested by Isaac Newton in *The Principia,* but he dismissed the phenomenon as being too small to measure. Almost presciently, he wrote, "Nay, whole mountains will not be sufficient to produce any sensible effect."[20] Nevertheless, in 1738, Pierre Bouguer, a French surveyor working in what is today Ecuador was able to measure the effect of a mountain's mass on a pendulum while working in the shadow of Mount Chimborazo.[21] Bouguer's measurement was rough, due to the harsh field conditions at the time, and he encouraged people to repeat his experiment under more favorable conditions. Even with his rough measurement, however, and only a vague idea of the size and mass of Chimborazo, Bouguer was able to determine that the interior of the earth must be significantly denser than the exterior—effectively disproving that the earth was hollow.[22]

One anomalous measurement can often be shrugged off, but in the 1760s, another set of surveyors working in another set of mountains also discovered that "down" wasn't exactly where they expected it to be. Beginning in 1763, Charles Mason and Jeremiah Dixon began drawing their eponymous line separating Pennsylvania from Maryland. The two colonies had been arguing over exactly where their border was for nearly one hundred years, and tensions were running high. However, when Mason and Dixon finished their work, they discovered that their line had wandered off course by almost nine hundred feet in some places. This figure was nearly twenty times the degree of error that they had been anticipating.[23] The error came about because their surveying equipment relied on a weighted line called a plumb bob to reliably measure angles relative to the surface of the earth.

In the 1760s, Pennsylvania and Maryland were both British colonies, and so Mason and Dixon's difficulties came to the attention of scientists back in England. When the chemist Henry Cavendish learned what had happened, he recognized that Mason and Dixon had stumbled on to the same phenomenon that Bouguer had noted in Ecuador. The mass of the Allegheny Mountains was affecting the plumb bobs in their equipment, causing their line to veer slightly off. But what Mason and Dixon saw as an irritating source of measurement error, Cavendish recognized as an opportunity. If you could find just the right mountain—one with a simple enough topography that you could precisely estimate its size and weight—you could use that mountain to weigh the earth itself.

In 1772, the Royal Society established the "Committee of Attraction," and tasked Charles Mason (sans Dixon) with finding a mountain that was suitable for determining the weight of the earth. He eventually settled on Schiehallion, a nearly perfectly conical mountain in the middle of an isolated plain in the center of Scotland. By 1776, the committee had not only chosen their mountain but also conducted sufficient surveying work and geological reconnaissance that they could estimate its size and density, and therefore its mass. When they measured the degree to which Schiehallion deflected a pendulum, they determined not just the mass of the earth but also that the earth had an average density of about 4.5 grams per cubic centimeter.[24] Over the course of his life, Cavendish worked on improving the instruments used to measure Schiehallion's gravity, and in 1798, he refined the estimate of earth's density to 5.448 grams per cubic centimeter.[25] That value is astoundingly close to the currently accepted value of 5.515 grams per cubic centimeter,[26] but it also posed a whole new problem.

The crust of the earth is primarily made up of two types of rock. Continents are made of granite, which has a density of 2.7 grams per cubic centimeter. The ocean floor is made of basalt, which has an average density of 3.0 grams per cubic centimeter. If the earth as a whole has an average density of 5.5 grams per cubic centimeter, then not only is the earth not hollow, but also whatever is inside must be significantly different from anything we see at the surface.

The Himalayas Are Big but Not That Heavy

Ironically, our first understanding of the properties of the earth's interior came from a study of one of its highest points, the Himalayan Plateau. The Great Trigonometrical Survey was a project intended to map the entirety of British Imperial territory in India as precisely as possible. Over the course of the endeavor, which spanned more than a century, oversight of the project passed between a series of superintendents, the second of whom was George Everest. Everest never came within sight of the mountain that now bears his name. In fact, he initially objected to the mountain being named for him.[27] The honor was bestowed despite his (possibly disingenuous) protests to commemorate the twenty years (from 1823 to 1843) that he had spent surveying India. This was a period of time during which he encountered a now-familiar problem.

Everest used two different methods for determining distances between towns in India. He mostly employed traditional land-based surveying techniques, but he double-checked those against astronomical sightings. In astronomical sightings, distances between points on the earth are determined by taking angular measurements from those points to a distant star. The technique requires a plumb bob, and of course, being so close to the gigantic mass of the Himalayan plateau, Everest's angles were off.[28]

Unlike Mason and Dixon, Everest's errors did not stay mysterious for long. Everest discussed his measurement problems with John Pratt, the Archdeacon of Calcutta (now Kolkata). Prior to arriving in India, Pratt had received his master's degree from Cambridge, where he studied gravity, and was the author of *Mathematical Principles of Mechanical Philosophy*.[29] Pratt was aware of Cavendish and the Committee on Attraction's work and realized that the gravitational pull of the Himalayas must be affecting Everest's measurements. When Pratt performed a rough calculation to try to verify his hypothesis, he made a shocking discovery: Everest's errors were far too small. A mass anywhere near the size of the Himalayan plateau should have been having a much larger effect than it was. For some reason, the Himalayas weighed far less than they should.[30]

Pratt's discovery of "missing mass" in the Himalayas led him to contemplate a fundamental question of geology: Why are there mountains in the first place? That is, why would some areas of the earth's crust stick up higher than others? His rather elegant solution was that the crust in mountain ranges must be less dense than other parts of the crust. Pratt knew from Cavendish's experiments that the interior of the earth was denser than the exterior.[31] Therefore, Pratt assumed that the crust of the earth must float on that denser interior. He reasoned that the crust in mountainous regions must be hotter, therefore less dense than the surrounding crust, and therefore floated higher.[32] Pratt's explanation for the missing mass in mountain ranges seems simple and elegant; yet it may actually have been more complicated than necessary.

Pratt presented his theory to the Royal Society of London in 1855, where it drew the attention of George Airy. Airy's official title was Astronomer Royal—a position once held by Edmond Halley—which made him not only the director of the Royal Observatory at Greenwich but also a member of the royal household. For all intents and purposes, he was in charge of time and space for the British Empire. When Airy read Pratt's explanation for the missing mass in the Himalayas, he published a reply, praising Pratt's extensive research and mathematical acumen, but also presenting his own simpler model to explain the phenomenon. According to Airy, you don't need the crust in mountain ranges to be warm to explain the missing mass; you only need the earth's interior to be fluid.

If whatever is beneath the crust is denser than the crust (as Cavendish demonstrated) and fluid (as Pratt implied), then the crust won't float on top of it like oil on top of water. The crust would float *in* that denser substance like a cork in water. According to Airy, mountains aren't places where the crust is hot. They are simply places where the crust is thick. The mass that is missing from mountain ranges isn't missing from the crust. It is missing because the thickened part of the crust projects down into the earth's interior, displacing the denser material underneath. (Figure 4.1 depicts the two models.)

Both Pratt and Airy created simple physical models that would explain not only the missing mass in mountain ranges but also the existence of mountain ranges in the first place. And there was a definite difference between the two models that could prove as an easy test. According to Airy, mountain ranges should have roots sinking deep into the inner layers of the earth. According to Pratt, the bottom of a mountain range should be at the same depth within the earth as the bottom of any other piece of crust. All geologists would need to do to determine who was right would be to find a way to look under the mountains. Unfortunately, that technology would take another sixty years to develop.

Looking through the Earth

Andrija Mohorovičić spent his life among the waves. He grew up in a coastal village in Croatia, where his father made anchors for a living, and his father-in-law was a sea captain. However, the waves that Mohorovičić himself devoted his life to were seismic waves.

Seismic waves are the waves generated by an earthquake. There are two basic types, with fairly self-explanatory names. Surface waves travel along the surface of the earth and do most of the damage in an earthquake. Body waves travel through the body of the earth and provide geologists with valuable information

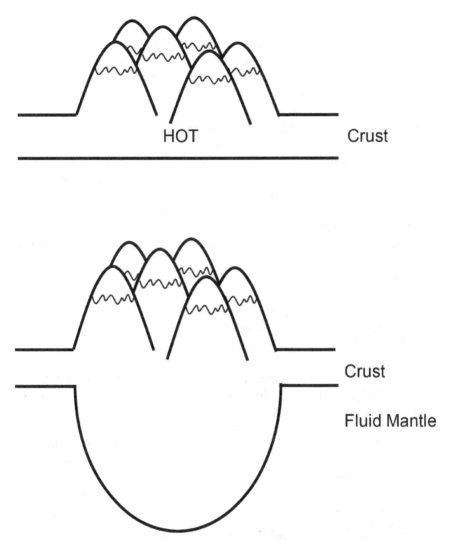

Figure 4.1. Two different models for why mountain ranges stick up above their surrounding crust. John Pratt suggested that the rock inside mountain ranges must be hotter than the surrounding crust. This would also create a density difference that allowed mountains to float higher on the mantle than surrounding cooler crust (top). George Airy suggested that if the mantle were fluid, then mountains didn't need to be hot. Any place where the crust was thick would float higher in the mantle like an iceberg in the ocean (bottom).

about what's inside. Body waves can be further divided into two types. Primary waves, or P-waves, move faster and are therefore detected earlier than secondary waves, or S-waves.

In addition to their speed, there is one other difference between P-waves and S-waves. P-waves are a type of compression wave similar to sound waves. Like sound waves, they move at a speed determined by the properties of the medium that they are traveling through. Scientists often talk about the speed of sound as though it were a constant, but it can vary considerably based on the temperature and density of the air it travels through. P-waves work the same way. The denser the material they are traveling through, the faster they go.

S-waves are translational waves, like the vibrations of a guitar string. They vibrate perpendicular to their direction of travel. This requires some amount of physical continuity to the medium they are traveling through. You can pluck a guitar string, but you can't pluck the air. S-waves work the same way and only travel through solid objects.

Whenever an earthquake occurs, it generates both body waves and surface waves. On October 8th, 1909, a significant earthquake struck the Kupa Valley in Croatia.[33] At the time of the Kupa earthquake, Mohorovičić was working as a lecturer at the University of Zagreb, only about forty miles away.[34] The earthquake was strong enough to destroy most of the buildings near its epicenter,[35] which made it an irresistible topic of study for Mohorovičić.

Mohorovičić gathered data from seismometers throughout the region and noticed a strange pattern in the P-wave data. The average speed at which the P-waves traveled away from the epicenter seemed to increase abruptly once you got more than about three hundred kilometers away. He realized that a sudden change in speed had to be the result of a sudden change in the medium that the waves were traveling through. Mohorovičić had discovered the bottom of the crust. By mapping out the path that the P-waves were traveling through the earth, he determined that the change in density happened at a depth of approximately fifty kilometers.

Today, the boundary between the bottom of the crust and the top of the underlying material (called the earth's mantle) is named the Mohorovičić Discontinuity. But that's a mouthful. If geologists are going to abbreviate *primary waves* to *P-waves* and *secondary waves* to *S-waves*, they certainly aren't going to say "Mohorovičić Discontinuity" over and over again. Its name is almost always abbreviated to just "the Moho."

The Moho separates the crust from the mantle. Above the Moho, the earth is made of two primary rock types. The continents are made of granite, which is light in color and relatively low in density. The ocean floor is made of basalt, which is darker in color and also a little bit denser. Under the Moho, the earth abruptly becomes denser. The rock in the mantle is called peridotite. It's a deep green color and much richer in metals like magnesium and iron than the crust is.

One of the most surprising things about the Moho is how close we are to it. Compared to the size of the planet as a whole, the crust is shockingly thin. There is some variability underneath the continents, but on average, the Moho there is about 38 km deep. Put another way, if you were to stand at the starting line of a marathon, you would actually be closer to the mantle than you would be to the finish line. The ocean crust is even thinner. Only 7 km of basalt separates the bottom of the ocean from the top of the mantle.

An important distinction to draw about the Moho is that it is a compositional boundary. It separates the aluminum-rich, low-density silicate rocks above it from the denser magnesium and iron-rich silicate rocks below. But that's not the only way to divide the earth into layers. Geologists also distinguish different layers of the earth by how they behave. The top 100 km of the mantle are rigid, like the crust above it. That combination of crust and rigid mantle is called the lithosphere. Once you get more than 100 km deep into the mantle, the combination of heat and pressure at those depths allows the solid rock of the lower mantle to behave like a fluid. This fluid solid layer is called the asthenosphere.

If *fluid solid* sounds like a contradiction in terms, that's understandable. The problem comes from the fact that people often use the words *fluid* and *liquid* interchangeably when, in fact, they mean different things. A fluid is anything that flows. Most liquids flow, but so do gases, so they both qualify as fluids. Under some circumstances, solids can flow too. Perhaps the best-known fluid solid is Silly Putty. Roll Silly Putty into a ball, and it stays in a ball. Drop that ball, and it bounces. These characteristics make Silly Putty a very obvious solid. But you can also stretch Silly Putty by applying gentle tension. Roll it into a log and hold that log by one side, and the other side will droop under its own weight and slowly flow downward under the pull of gravity, just like any other fluid would.

The confusion that arises from the imprecise use of language is another of the Idols of the Mind that Francis Bacon warned of in *Novum Organum*: the Idol of the Marketplace.[36] The name comes from the fact that it only comes about when people interact with one another. It's hard to have a miscommunication with yourself. This is also a very subtle problem. It might not always be obvious when two people are using words differently from one another. But it can be a particularly vexing barrier to understanding the world around us.

By discovering that seismic waves could be used to map regions of high and low density inside the earth, Mohorovičić gave geologists the tool necessary to resolve the debate between Airy and Pratt about what is happening under mountain ranges. Neither was exactly right, but Airy was closer. Mountain ranges are places where the crust is thicker than the surrounding area, but they don't just float in the mantle. Instead, the mass of a mountain range bends the lithosphere, causing the entire region to dip down into the asthenosphere. This model of regional compensation was originally proposed in the 1930s by Dutch geophysicist Felix Vening Meinesz to explain not only the presence of mountains but also why so many mountain ranges are accompanied by low-lying basins around their edges.[37]

Seismic wave data lets us know that the crust is thin. Conversely, the mantle is very thick. It is most of the earth, about 84 percent by volume. But that presents a new problem. Although the mantle is denser than the crust, it's not that much denser. The density of the mantle changes with pressure and depth. Near the top, the mantle has a density of 3.3 grams per cubic centimeter. That number increases with depth to a maximum of about 5.6 grams per cubic centimeter at the bottom of the mantle. Overall, the mantle has an average density of about 4.4 grams per cubic centimeter. And there's the problem: If 84 percent of the earth is significantly less dense than the earth as a whole, then whatever is left must be incredibly dense.

Is the Core Liquid or Solid?

When an earthquake occurs, it generates both P-waves and S-waves that travel through the solid body of the earth. The speed of a P-wave is determined by the density of the substance it is traveling through, and therefore P-waves accelerate when they hit the Moho and move from the crust to the mantle. Then 2,900 km later, they speed up again when they cross the boundary between the mantle and the innermost layer of the earth, the core. That lets us know that the core is even denser than the mantle.

S-waves also travel through the earth. They only travel through solids, so the fact that they travel through the mantle confirms that even though the asthenosphere is fluid, it is a fluid solid. But S-waves don't speed up when they cross the core-mantle boundary. They stop. The layer of the earth directly below the mantle isn't just dense. It is a dense liquid.

Even before we discovered how to use seismic waves to look through the earth, geologists had good reason to suspect that the earth contained a dense iron core. Outside of science fiction novels, we have never been to our own planet's center, but we have observed the deep interiors of other planets—through meteorites. Meteorites are leftover pieces from the formation of our solar system. Some are pieces of asteroids and planets that collided and broke apart early on. Others are the actual unused building blocks from which the planets formed. Importantly, meteorites come in two broad types. Most meteorites are stony meteorites composed of silicate minerals like those found in our crust and mantle. Non-stony meteorites are made of iron.[38] So a good first guess for what a planet like ours is made of would be mostly silicate minerals, with a smaller amount of iron.

Based on its density, the core is in fact a large chunk of mostly iron sitting at the center of our planet. This broad description of the earth as a whole was first proposed in 1896 by German physicist Emil Weichert, but he believed the entire earth was solid.[39] The fact that S-waves stop when they hit the core-mantle boundary suggests otherwise, but there was considerable resistance to the idea of a liquid core until well into the 1920s.[40] The resistance was partially because it just didn't seem possible for iron to melt that deep inside the earth.

When Cyrus Teed put the sun in the center of the earth, he may actually have been underestimating how hot it is down there. The surface temperature of the sun is about 6,000 degrees Celsius. Estimates of the temperature in the core vary, but they range up to 7,000 degrees Celsius.[41] Ordinarily, iron melts at about 1,500 degrees Celsius, so it might be hard to imagine why anyone would ever think the core could be solid. But, under the pressures that exist at the center of the earth, that melting point rises to an unknowable degree. So, prior to S-wave data, a solid core was perfectly plausible. However, in 1926, geophysicist Harold Jeffreys used the newly available seismic data to demonstrate that the earth's center was liquid. This would be considered the final word on the center of the earth—for about ten years.

Inge Lehmann, born in Copenhagen in 1888, received her initial education at a new school in Copenhagen called the Fællesskolen or *shared school*. The school was unique for the time in that it made no distinction between boys and girls, teaching them the same subjects with the same rigor, and even having them play sports together.[42] Lehmann left the school with a love for mathematics and an assumption that men and women were treated equally everywhere. This assumption, she would later reflect, set her up for some disappointing realities.[43]

Lehmann moved to England in 1911 to spend a year at the University of Cambridge but was shocked to discover that she did not have the same freedoms and opportunities there as the male students. After less than a year at Cambridge, Lehmann found herself so exhausted by the stifling environment that she became physically very ill, left school, and left academia altogether.[44] She spent the next several years doing actuarial work for an insurance company back in Denmark, where she earned a reputation for incredibly precise data work.

Lehmann eventually finished her mathematics degree and found a position as assistant to a professor of actuarial sciences at the University of Copenhagen. A few years later, she was transferred to a different department in the university and became an assistant to a professor of seismology. While working for him, she discovered her own love of the field. This led Lehmann to pursue her degree in seismology and to eventually succeed her mentor as director of seismology for the Royal Danish Geodetic Institute in 1928.[45] Lehmann's position at the

Geodetic Institute involved maintaining some of the northernmost seismometers in the world, in both Denmark and Greenland. These are remote locations but would prove to be ideally placed to gather important seismic data.

On the morning of June 17, 1929, a magnitude 7 earthquake struck just north of Murchison, New Zealand. The earthquake destroyed buildings, collapsed coal mines, triggered landslides, and killed seventeen people. Body waves from the earthquake registered on seismometers all over the world. Denmark is at almost the exact opposite point on earth from New Zealand, so the P-waves that registered there traveled directly through the centermost parts of the earth's core. They just didn't travel in exactly the way seismologists would have expected.

Like Mohorovičić before her, Lehmann discovered a discontinuity. P-waves traveling through the very center of the earth were being refracted slightly by a surface between two substances with slightly different properties. This was a very small refraction, and its discovery required meticulous data analysis over many years. In 1936, Lehmann published a paper titled simply "P," in which she made the case that the Lehmann Discontinuity (as it would eventually be named) marked the boundary between a solid inner core and a liquid outer core.[46] Its discovery was a crucial step in finally solving Edmond Halley's mystery of the moving north pole.

Why does the magnetic north pole move? For that matter, why does the earth have a magnetic north pole to begin with? An obvious answer would be that the solid iron of the inner core is magnetic. But that can't be right because magnets lose their magnetism when heated above 770 degrees Celsius, a temperature called the Curie Point. At the extreme temperatures we find in the core, nothing could possibly be magnetic.

The exact mechanism by which the earth generates its magnetic field is still only partially understood, but the best candidate appears to be the outer core. The outer core is a liquid layer of the earth sandwiched between two solid layers. What's more, the layer above it, the mantle, is much cooler than the inner core below it. As such, the liquid in the outer core must be moving. The liquid iron at the bottom of the outer core is being heated by the inner core, rising through the rest of the outer core, and then sinking again when it gets cooled by the base of the mantle. This pattern of fluid motion due to heating from below is called *convection*.

At the same time that the liquid iron of the outer core convects, it also spins along with the rest of the earth. Geologists who study the earth's magnetic field (called *geodynamicists*) believe that it is this complex motion of the electrically conductive iron in the outer core that somehow generates the magnetic field. As a general rule, scientists hate the word *somehow*, but at the moment, that is where our models of the earth's magnetic field remain.

The problem is that with all of the different forces, heat sources, and rotations acting on it, the motion of the outer core is incredibly complex. Add to that complexity the fact that the magnetic field itself can affect the motion of the fluid iron generating it, and things get downright chaotic.[47] Our best mathematical models for how the outer core moves still don't perfectly describe the magnetic field as we actually see it. However, the models come close in a lot of ways, including predicting a field with a north and a south pole that are close to the geographic poles but wander slightly over time.[48]

Ironically, the time Lehmann spent as an actuary was probably invaluable to her discovery of the inner core. Her nephew Niles Groes once described watching her work to understand the implications of her P-wave data and seeing her organizing her data in oatmeal boxes very similar to the filing system she had used in her actuarial work.[49] In 1971, when Lehmann was being awarded the American Geophysical Union's highest honor, the Bowie Medal, her citationist said that the Lehmann Discontinuity "was discovered through exacting scrutiny of seismic records by a master of a black art for which no amount of computerisation is likely to be a complete substitute."[50] Lehmann saw things from a slightly different perspective. When asked about her work ethic and her reputation for meticulous data work, she remarked, "You should know how many incompetent men I had to compete with—in vain."[51]

The Bowie Medal has been given out annually since 1939. Among its first sixty recipients, Inge Lehmann was the only woman.[52] Lehmann maintained her post as director of seismology for the Royal Danish Geodetic Institute for twenty-five years, including through the Nazi occupation of Denmark, until her retirement in 1953. However, she kept working in seismology and geophysics until the age of ninety-nine when she published her last paper. She died in 1993, just a little shy of her 105th birthday. Women are still underrepresented in geophysics today. Since 1997, the American Geophysical Union has awarded the Inge Lehmann Medal to recognize excellence in the study of the earth's interior and to honor Lehmann's legacy. As of this writing (January 2022), only one woman, Dr. Barbara Romanowicz of the University of California at Berkeley, has ever won it.[53]

CHAPTER 5

The Earth Is Old

It takes about eleven days to count to 1 million. That's assuming that you count at the rate of one number per second. Of course, you could probably count "one, two, three . . ." faster than that, but for every number with a name like *one*, there are several more with names like *seven hundred ninety-six thousand eight hundred seventy-four*. So one number per second is probably a fair, or even generous, estimate of average counting time. Billions are even bigger than millions. Billions are bigger than millions in a way that most people have trouble grasping. Counting to a billion takes nearly thirty-two years. According to our best current estimates, the earth is 4.5 billion years old. Counting to that number would be an intergenerational undertaking, requiring more than 142 years to complete.

The numbers that geologists use when talking about the age of the earth, or the antiquity of geological events, are so large that it is hard to imagine where they come from, or how they could possibly be used precisely. If the earth is really 4.5 billion years old, can we truly be certain that it's not 4.4 or 4.6? Geologists regularly publish dates for past events with error bars like, "plus or minus 250,000 years."[1] How did we get to the point where the entire age of the human species became a rounding error? We didn't start this way. We started out measuring the age of the earth in days.

Attempts to Date Biblical Creation

One of the first scientific texts to put a definitive numerical date on the origin of the earth was published in 1643. Thomas Browne's *Religio Medici* (*The Religion of a Doctor*) wasn't a geology or astronomy book. It was a reflection on the relationship between science and religion by a Renaissance-era doctor—a man who needed, at times, to live in both of those worlds. In his book, Browne contemplates the difference between time and eternity stating that "Time we may comprehend, 'tis but five days elder than ourselves."[2] Browne was referring to the cosmogenic chronology laid out in the Book of Genesis. In this framework, the age of the earth, and in fact the universe, is easy to determine. The universe, including the earth, was created on day 1, and we were created on day 6.

Browne's comment on the age of the earth wasn't intended to have any real geological merit. It was part of a commentary on what the human mind can and cannot comprehend.[3] However, a few years after Browne, an Irish clergyman named James Ussher used that same Biblical logic in an explicit attempt to pinpoint the date of creation. If the creation of the earth occurred just five days before the creation of the human species, then all you would need to do is determine how long ago the first human was created, and you could, in fact, determine the age of the earth.

The data you would need to determine the date of the first humans is partially provided in the Book of Genesis chapter 5, in what is sometimes referred to as the "list of begats." "And Adam lived a hundred and thirty years, and begat . . . Seth . . . and Seth lived a hundred and five years, and begat Enos . . . and Enos lived ninety years, and begat Cainan . . ." and so forth.

Ussher wasn't the first person to try to use information from the Bible to date the creation, but his efforts are perhaps the best known and the most maligned in the history of science. They are also often dismissed as a simple act of addition. Add up all the begats and you get Adam's birthday, and therefore the age of the earth. It's actually more complicated than that. That list of begats only gets you so far and then stops. Ussher was no mere accountant. To get an actual age of the earth, you need to compare multiple biblical lineages against the Babylonian and Roman historical records.[4] In 1654, after several years of work, Ussher determined that the lineage of Old Testament patriarchs, and therefore the universe, itself began in 4004 BCE.[5] The 1600s and 1700s were the Golden Age of horribly long book titles; and so, Ussher published his assessment of the earth's age in a book called, *Annales Veteris Testamenti, a Prima Mundi Origine Deducti, una cum Rerum Asiaticarum et Aegyptiacarum Chronico, a Temporis Historici Principio usque ad Maccabaicorum initia Producto* (or, mercifully, just *The Annales*).

If he had stopped at just 4004 BCE, then perhaps James Ussher would just be one more Renaissance theologian that nobody today had ever heard of. But Ussher went further, declaring the creation to have begun on Sunday, October 23, 4004 BCE, with light being created around midday.[6] This degree of precision is impressive, and based on an extensive study of Biblical and Roman calendrics, but rings ridiculous to modern ears. As a character in the 1955 play *Inherit the Wind* puts it "That Eastern Standard Time? . . . It wasn't Daylight Saving Time, was it? Because the Lord didn't make the sun until the fourth day!"[7]

Although Ussher has spent the last few hundred years as the subject of mockery by the scientific community, his result was completely consistent with contemporary Biblical scholars working on the same problem. Most theologians trying to date the creation came up with about a six-thousand-year-old earth. As it turns out, scientists working from non-historical and non-Biblical data in Ussher's day were making similar assumptions.

Dating the Earth by Dating the Oceans

Before going any further on the age of the earth, I feel compelled to briefly sing the praises of Edmond Halley. Halley's contributions to both astronomy and physics are hard to overstate. In addition to establishing the periodicity of comets, he determined the distance between the sun and the earth, did extensive mapping of the stars, and even helped to fund the publication of Newton's *Principia*.[8] He just seems to have had bad luck when it came to geology. In 1715, in between developing his hollow earth theory and attributing the aurora borealis to the light of an internal sun, Halley turned his attention to the age of the earth.

Halley started his investigation into the age of the earth with a study of terminal lakes. A terminal lake is a lake that has rivers flowing into it, but no rivers flowing out of it. A classic example of a terminal lake would be Great Salt Lake in Utah, which is fed by the Jordan, Bear, and Weber rivers, but which has no rivers flowing out from it. Halley realized that since the water evaporating out of those lakes is pure water, any dissolved minerals brought into those lakes by rivers would be left behind and concentrated.[9] Those lakes would start out as freshwater lakes but, over time, would have continuously increasing salinity levels.

Halley correctly realized that the world's oceans must have accumulated their saltiness through a similar process. Geologists recognize that the oceans began as freshwater and accumulated their current salt content through the accumulation of dissolved ions brought into them by rivers eroding continental rock. Halley suggested that if we could determine the amount of salt delivered into the oceans by rivers in a given year, and if we could measure the current saltiness of the oceans, then we could determine how long that salt had been accumulating.[10]

Of course, Halley also realized that determining the amount of salt being delivered into the world oceans would require measuring the salinity and flow rate of every single river on earth. That would be a herculean task today and an impossible one in 1714. But he did have a workaround that both shows his ingenuity as a scientist and hints at the answer he was expecting to get. Halley suggested that if we could find an ancient Greek or Roman text that gave the

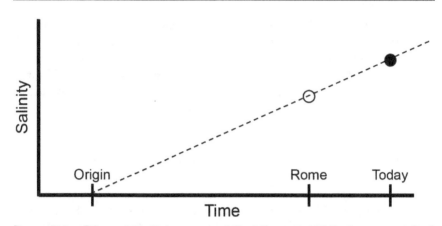

Figure 5.1. Edmond Halley proposed that if we could find some ancient Roman text that included a measured salinity of the ocean, we could date the age of the earth. Comparing the salinity today (solid dot) to the salinity in Roman times (hollow dot) would provide a rate of change that could then be extrapolated back to determine how long the oceans had been gaining salt (dotted line). This method makes the assumption that the amount of time that has passed since the days of the Roman Empire is a significant enough percentage of earth history that there would be a measurable change over that time period.

salinity of the oceans in that age, then we could use that age to extrapolate a rate of change.[11] (See figure 5.1.)

Nobody ever found such a document, but the fact that Halley thought this extrapolation method would work shows approximately how old Halley thought the world was. If he was expecting to find a measurable change in the salinity of the oceans between 1714 and the Roman Empire, then he must have believed that the intervening time period was a significant portion of the age of the earth. Halley's method never worked, but just from his proposal, we can tell that he must have expected the age of the earth to be somewhere in the tens of thousands of years at most.

Halley's "salt clock" model, while unsuccessful, is enlightening in that it illustrates the information necessary to answer a question like "how old is the earth?" All clocks, from sundials to kitchen timers to Halley's salt clock, work in the same way: they combine a measurable change with a known rate of change. If you know how much something has changed (the shadow of a sundial, the hands of a clock, the salinity of the ocean) and you know how long it takes for that change to happen, then you know how much time has passed.

The salt clock was not a bad idea. It was just untenable. In 1899, the Irish physicist John Joly once again attempted to date the age of the oceans using the accumulation of salt. But Joly's result gave an earth that was hundreds of millions of years old.[12] Obviously, this increase in age is only partially the result of better data. Making the leap from thousands to nearly billions requires an entirely different conceptual framework for thinking about the age of the earth. Somewhere in the 185 years between Halley and Joly, there must have been a major change not just in data or in method, but in how we think that the earth works in general.

From Thousands of Years to Limitless Time

James Hutton was born in Edinburgh in 1726. He is alternately referred to as either the founder or father of modern geology. His pathway to becoming the first modern geologist was a winding one. At various points in his life, Hutton was a failed lawyer, a medical student who wrote his dissertation on the circulation of the blood, an industrial chemist, and a gentleman farmer. He was also an avowed deist.[13]

Deism was a seventeenth- and eighteenth-century movement that spanned the gap between a religion and a philosophy. It believed that studying nature empirically could tell us about the nature of God, but it also believed that God was rather remote. According to deism, God is good, but His goodness is manifested in the fact that He set up knowable and generally benevolent laws in the universe. Since God was a truly competent creator, He didn't need to intervene in our daily lives. The simple, mechanical workings of the universe would inevitably cause His plan to unfold according to His wishes.[14]

Hutton had inherited two farms from his father and moved to one of them after making his own fortune selling chemicals to dye-makers. For Hutton, farming was an intellectual pursuit. He learned the theory and best practices for farming and let someone else do the physical labor. He studied with the best agronomists in Scotland, but whenever he bought a new plow, he made sure to get a new plowman to go with it.[15]

Hutton's deistic tendency to see all of nature as a reflection of God's goodness combined with his academic interest in farming led him to notice a conundrum that he called the Paradox of the Soil. Soil is essential not just for farming but for all life on earth. However, soil comes from the weathering and erosion of the continents. Even a quick glance at a map is all that it takes to realize that there is far more ocean than continent on earth. Erosion left unchecked will eventually and inevitably reduce the continents to nothing.[16] For a deist, the Paradox of the Soil is deeply troubling. Why would God create a process that is essential to life on earth, but that will ultimately doom life on earth?

The seeds for resolving the Paradox of the Soil were planted years before Hutton ever turned his attention to planting. When he received his medical degree in 1749, Hutton wrote his thesis on the circulation of the blood. He marveled at the balance of the circulatory system. Blood was constantly being consumed as it nourished the body, but it was also being constantly replenished.[17] If that was true for blood, then it might also be true for the continents. Somewhere in the earth, Hutton reasoned, there must be some mechanism for replenishing the continents.

In 1795, James Hutton published *Theory of the Earth*. This four-volume treatise resolved the Paradox of the Soil while simultaneously demonstrating the igneous nature of granite, proposing one of the first theories of how mountains are formed, and laying the theoretical framework for all of modern geology. It was also completely incomprehensible to most people. For all of Hutton's gifts and talents, he could not write a straightforward sentence to save his own life. Here is Hutton questioning whether or not the earth is solid:

> This first part is commonly supposed to be solid and inert; but such a conclusion is only mere conjecture; and we shall afterwards find occasion, perhaps, to form another judgment in relation to this subject, after we have examined strictly, upon scientific principles, what appears upon the surface, and have formed conclusions concerning that which must have been transacted in some more central part.[18]

Here is Hutton saying that the earth moves:

> There is the progressive force, or moving power, by which this planetary body, if solely actuated, would depart continually from the path which it now pursues, and thus be for ever removed from its end, whether as a planetary body, or as a globe sustaining plants and animals, which may be termed a living world.[19]

And here is Hutton saying that the earth is old:

> Time, which measures every thing in our idea, and is often deficient to our schemes, is to nature endless and as nothing; it cannot limit that by which alone it had existence; and, as the natural course of time, which to us seems infinite, cannot be bounded by any operation that may have an end, the progress of things upon this globe, that is, the course of nature, cannot be limited by time, which must proceed in a continual succession.[20]

Hutton's prose was so nightmarish that for seventeen years, nobody really had any idea what he was talking about. He presented his theory of the earth as an address to the royal society in 1785.[21] He published *Theory of the Earth* as a book in 1795.[22] But it wasn't until 1802, when his friend John Playfair wrote *Illustrations of the Huttonian Theory of the Earth*[23]—essentially translating *Theory of the Earth* from Hutton into English—that anyone really understood what his model said.

Being both a good deist and a good Enlightenment Scotsman, Hutton called his model the World Machine. It's an interesting model, but it happens to be very wrong. The model begins with mountains that are eroding, creating sediment that eventually flows into ocean basins. When enough sediment accumulates in the ocean, the mass of that sediment creates an instability in the earth, causing the ocean basins to be violently thrust upward. Exactly how accumulated mass leads to uplift is unclear, but this was Hutton's claim. These uplifted areas become new mountains, resulting in mountains where the oceans used to be and oceans where the mountains used to be. From this point, the cycle begins again.[24]

By creating a mechanism by which new mountains form, the World Machine resolves the Paradox of the Soil. The damage done to the continents by erosion and soil formation would eventually lead to the uplift of new mountains, which would then be eroded creating new soil and eventually new mountains once again. In theory, the world could now go on forever. But there was a second assertion in *Theory of the Earth* that would fundamentally transform geology as a science. Hutton didn't merely propose the uplift of new mountains as a possible future event. He proposed that this cycle of erosion and uplift had already played out in the past—an unknowable number of times, over an unknowable span of time.

Siccar Point is a narrow finger of rock jutting into the North Sea a few miles north of the border between England and Scotland. On a boat ride around the point, in 1788, Hutton spotted a rock outcrop where two different rock types met at an unusual angle.[25] Hutton knew that sedimentary rocks, since they form on the seafloor, typically form as horizontal beds. But here were beds of grey sedimentary rock (called greywacke) standing vertically on end. Above them were horizontal layers of a red sandstone. Where the two rock types met, there was an obvious plane of contact where the greywacke had been partially eroded away.

Hutton interpreted the geometry of Siccar Point as evidence that the World Machine had run in the past. In his view, the greywacke had formed horizontally on the floor of the ocean. They were then violently thrust upward, resulting in their current orientation. After a period of time during which the greywacke was being eroded away, this area must have again been underwater so that the sandstone could form. Finally, the fact that all of these rocks were above water and eroding again indicated a second period of uplift.[26]

According to Hutton, all of this geologic history must have taken time—beyond thousands of years. Just a few miles from Siccar Point are the ruins of St. Helen's Church. St. Helen's was built in the thirteenth century from the same red sandstone exposed at Siccar Point and still had an entire wall standing. Just one hundred miles to the south was Hadrian's Wall—nearly two thousand years old and still standing. Hutton realized that if stone walls and stone churches can last for centuries and millennia, but entire mountain ranges had eroded away, then the earth must be incomprehensibly old.[27] *Theory of the Earth* famously ends, "The result, therefore, of our present enquiry is, that we find no vestige of a beginning,—no prospect of an end."[28] In Hutton's estimation, not only had he resolved the problem of erosion eventually destroying the earth, but he had also eliminated any constraints on how far back into the past the history of our world might go.

It is again worth noting that the World Machine is not actually how the earth works. As we'll see in a later chapter, Hutton wasn't even remotely close to a correct explanation of how mountains form. However, his insights that the earth has a complex geological history and that creating its current landscape took vast amounts of time were spot on. *Theory of the Earth* moved geologists from thinking about the origin of the earth as a fairly recent event that happened several thousands of years ago to a more distant event that happened several millions of years ago.

The problem was, how many millions of years? For the next several decades, geologists would propose ages of the earth ranging from 20 million years to 200 million years, which were just best guesses. There wouldn't be an age of the earth based on actual empirical measurement until someone changed the question slightly.

Measurable Change and a Known Rate of Change

At about the same time that Hutton was giving geologists license to think about the history of the earth in millions of years, astronomers were speculating on how the earth formed in the first place. The most common opinion, expounded by both philosophers like Immanuel Kant and astronomers like Pierre-Simon Laplace, was that the earth, and the entire solar system for that matter, started out hot.[29] To this day, geologists believe that the earth began as a fiery ball of molten rock sitting in the cold vacuum of space. This model provides a new way to think about the age of the earth. Instead of asking "How long has the earth existed?" we can ask "How long has the earth been cooling?" Rephrasing the question this way changes the age of the earth from a geology problem to a physics problem.

In 1862, the British physicist William Thomson tried to calculate the cooling time of the earth. In the 1860s, if you had a physics problem, especially one related to heating and cooling, there may have been no better person to bring it to than Thomson. Thomson spent his entire career at the University of Glasgow working on thermodynamics—the study of how thermal energy moves and behaves. His work was so groundbreaking that in 1892, Thomson became the first British scientist to be ennobled for his work.[30] He became the First Baron Kelvin, named for the River Kelvin that flowed near the University of Glasgow. The Kelvin scale of absolute temperature would later get its name in honor of the newly minted Lord Kelvin (the name by which Thomson is better known today).[31]

To understand Kelvin's model of how the earth cooled, it helps to imagine the earth as being like a large chicken potpie. When a chicken potpie first comes out of the oven, both the crust and the filling are extremely hot. The crust cools quickly, but the filling can stay hot for much longer. Many people learn this lesson painfully by burning their mouths on a potpie that seems cool on the outside but is still very hot on the inside. Eventually, the inside also becomes cool enough to eat, but that can take a bit longer. Therefore, you can tell how long a potpie has been cooling by the difference in temperature between its outermost crust and the gooey interior immediately underneath the crust.

Kelvin didn't use any culinary metaphors, but he did propose that the earth started hot and that its outermost crust may have cooled very quickly. He therefore believed that if you could measure the changes in temperature that accompany changes in depth (which geologists call the *geothermal gradient*), you could determine how long the earth has been cooling.[32] By 1862, there had been enough deep mines dug around the world, that Kelvin was able to get a rough estimate of the geothermal gradient of about 1° F per fifty feet of depth and used that estimate to determine that the earth had been cooling for somewhere between 20 million and 400 million years with a most likely age of approximately 98 million years.[33]

Considering the fact that Kelvin had apparently solved one of the biggest questions in geology, you would think that geologists would have been delighted by his results. But the general reception of Kelvin's heat-based date was lukewarm at best. Geologists had been busy in the decades that had passed since Hutton's *Theory of the Earth*, and all of their contemporary models of how the earth worked allowed, or even required, immense expanses of time. As an example, Charles Darwin had published *On the Origin of Species*, presenting a model of evolution that by his reckoning required hundreds of millions of years to work.[34] Geologists had only been thinking on long time scales for a few generations, but those scales had already become indispensable. And Kelvin's age of the earth seemed distressingly low.

Perhaps the most direct attack of Kelvin's calculation came from the English naturalist Thomas Henry Huxley, who simply dismissed the result, saying, "True or fictitious, they have made no practical difference to the earth, during the period of which a record is preserved in stratified deposits."[35] Given his choice between trusting in Kelvin's estimates or trusting in the evidence of the rock record, Huxley chose the latter. If mathematics disagreed with geology, then the problem must be with the math. Huxley wrote:

> Mathematics may be compared to a mill of exquisite workmanship, which grinds you stuff of any degree of fineness; but, nevertheless, what you get out depends on what you put in; and as the grandest mill in the world will not extract wheat-flour from peascods, so pages of formulæ will not get a definite result out of loose data.[36]

Huxley's willingness to reject physics in favor of geology seems almost heretical now. That is because we have tacitly given physics a special place in science. There is an unspoken hierarchy in our perceptions of the sciences that privileges those "hard sciences" in which the subjects of study can be more easily quantified. This hierarchy has led to a phenomenon in the sciences called *physics envy*, where other sciences try too hard to quantify their data in an attempt to seem more "scientific."[37] Physics envy is another example of the argument from authority logical fallacy, in which we believe an argument not on its own merits but because we trust and respect the source making the argument.[38] Huxley may not have shared our modern awe for the authority of physics, but Kelvin most certainly did. The reticence that geologists had with adopting his date seems to have truly baffled him. He sometimes went so far as to accuse geologists of willful ignorance. He once reportedly told the geologist Andrew Ramsey, "You can understand the physicists' reasoning perfectly if you give your mind to it."[39]

In 1866, Kelvin published a second paper on the age of the earth, refining his date based on new measurements of the geothermal gradient and revised estimates of the earth's internal temperature. This time around, he said that the absolute oldest the earth could possibly be was 20 million years old and that even that age was probably an overestimate.[40] Kelvin was well aware by then that geologists would be outraged and say that the date was impossible because it did not allow enough time for the earth to work as they believed it did. He preempted that argument in the title of the paper. If 20 million years wasn't enough time for geologists to account for the rock record, then everything in geology must be wrong. The paper was titled "The 'Doctrine of Uniformity' in Geology Briefly Refuted," and it was a single 203-word paragraph—almost exactly the length of this paragraph you're reading. It was the scientific equivalent of a mic drop. And it would remain the definitive word on the age of the earth for almost another forty years. After all, how do you dispute physics, let alone Lord Kelvin himself?

Radioactivity and the Age of the Earth

In 1904, at the age of eighty, Lord Kelvin attended a lecture at the Royal Institution by a physicist from New Zealand named Ernest Rutherford about an exciting new discovery that was reshaping our understanding of nature—radioactivity.[41] Radioactivity had only been discovered a few years earlier by the French physicist Henri Becquerel, but it was one of the most exciting discoveries in the physics community in generations. Of the six recipients of the newly created Nobel Prize in physics, four of them had won for discoveries related to radioactivity.[42] Rutherford (who would win his own Nobel Prize in 1908) was there not only to lecture on radioactivity in general but also to explain how it changed our estimates of the age of the earth.

Understanding how radioactivity works requires understanding the structure of an atom. Every atom in the universe has two fundamental parts: the nucleus and the electron cloud. An electron cloud is a strange place, full of nearly massless particles called electrons moving at near light speed. It is the portion of the atom that determines how each atom bonds to other atoms. It is difficult to understand the behavior of electrons without diving into a world of quantum theory and complex mathematics. It is also somewhat tangential to the subject of radioactivity. So we'll ignore it.

The nucleus of an atom is where almost all of its mass is located. Some of that mass comes in the forms of large, positively charged particles called protons. *Large* is, of course, relative. A line of 90 billion protons lined up neatly in a row would be about as long as the average human hair is wide. But that still makes protons about one thousand times larger than electrons. Protons also let you know which chemical element the atom represents. If an atom has one proton in its nucleus, regardless of anything else that might be happening in the rest of the atom, that is an atom of hydrogen. Two protons make helium. Three protons make lithium. And so forth, down the periodic table.

The fact that protons are positively charged means you can't have an atomic nucleus with just two protons in it. The two like-charged particles would repel one another and blow the nucleus apart. Fortunately, there is another kind of particle in the nucleus—the neutron. Neutrons are about as heavy as protons. They also have a force associated with them that is more than one hundred times stronger than electromagnetism and that holds the nucleus together. Physicists, being imaginative types, call this the *strong nuclear force*.

Being a hydrogen atom means you only have one proton in your nucleus. But it doesn't necessarily tell you how many neutrons are in your nucleus. Hydrogen atoms exist with zero, one, or two neutrons in the nucleus. These different varieties of hydrogen are called *isotopes*. Isotopes are distinguished from one another by the total number of particles in their nuclei, so hydrogen with one proton and no neutrons is called Hydrogen-1 (or more typically ^{1}H). Hydrogen atoms with 1 or 2 neutrons are called ^{2}H and ^{3}H, respectively. They are all hydrogen. So, for example, any of them can be the H in H_2O. But there are also going to be distinct differences in the way they behave out there in nature.

One property that varies from isotope to isotope is radioactivity. Simply put, radioactivity is the tendency of some nuclei to spontaneously break apart. The atom will expel or transform a small chunk of its nucleus in a process called *radioactive decay*. After decay happens, the remaining atom will be a different isotope of a different element. For example, Uranium-235 can decay into Thorium-231. This makes ^{231}Th the *daughter product* of ^{235}U. Technically, any isotope can decay. Some just do it more often than others. An atom of iron will likely stay iron until the end of the universe. On the other hand, some cobalt atoms are lucky to last a week.

Physicists describe how radioactive an isotope is by describing the isotope's *half-life*—a term that can be taken quite literally. A half-life is the amount of time it takes on average for half of a radioactive isotope to decay to its daughter product. Counting individual atoms is difficult. But counting ratios of different atoms in a substance can be done fairly easily in the lab. If a rock contains some amount of a radioactive isotope when it forms, then after one half-life of that isotope has passed, half of the isotope should have decayed to daughter product. That means the ratio of isotope to daughter product should be 1:1. Once another half-life passes, half of the remaining isotope will have decayed, leaving only 25 percent of the original amount behind and creating a 1:3 ratio of isotope to daughter product. And so on through the ages. (See figure 5.2.)

Rutherford would eventually publish the first radioactivity-based estimates of the age of the earth. In 1905, he would use uranium isotopes to demonstrate that the rock samples he was studying (and therefore the earth itself) must be at least 140 million years old.[43] In 1906, he would revise that date up to 500 million years based on newer findings.[44] While these estimates had not yet been published when Rutherford gave his address to the Royal Institution in 1904, the audience in attendance (including Lord Kelvin) knew what Rutherford had been working on, and knew that his results contradicted Kelvin's age of the earth.

The circumstances of Rutherford's address had all the makings of a classic showdown. The venerable old hand was there to defend his reputation from the hot young upstart. Years later, Rutherford would tell a biographer that he felt terribly nervous as the lecture began, seeing Kelvin in the room and knowing that the entire second half of his address contradicted Kelvin's work[45]; but then, Kelvin did something completely unexpected. He took his seat in the lecture hall and almost immediately fell fast asleep. As Rutherford reached his conclusion about how radioactivity could be used to demonstrate that the earth was, in fact, at least ten times older than Kelvin had estimated, Kelvin woke up. As Ruther-ford put it, "I saw the old bird sit up, open an eye and cock a baleful glance at me!"[46] At this point, Rutherford added an impromptu line to his planned ad-dress that would haunt geology for nearly a century.

Rutherford's improvisation was to declare that his findings did not actually contradict Kelvin at all. Kelvin's model, according to Rutherford, assumed that all of the earth's internal heat was primordial; that is, that the earth had only been cooling over time with no sources adding new heat into the mix. Since ra-dioactivity produces heat, Kelvin could not have known that his model would be off. If anything, his underestimates of the earth's age prophetically demonstrated that our understanding of the earth was incomplete.

Rutherford's *ad hoc* explanation—that the previously unknown heat gener-ated by radioactivity explained Kelvin's error—was very clever. It also appeased Kelvin, who smiled at Rutherford and then went back to sleep.[47] Unfortunately, it was also incredibly wrong. Rutherford knew that Kelvin's heat-based estimates for the age of the earth were off by at least a factor of 10, possibly a factor of 100. There is radioactive heat being generated in the earth, but not nearly enough to make the earth look that young. Despite its physical impossibility, the explana-tion persisted for decades.

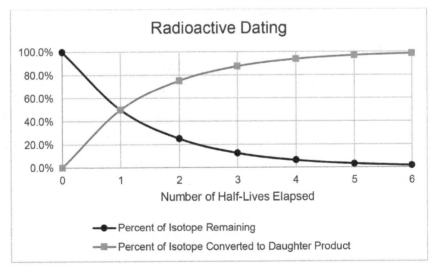

Figure 5.2. As time elapses, the amount of radioactive isotope in a rock sample decreases and the amount of daughter product increases. The ra-tio of isotope to daughter product lets you know how many half-lives have elapsed since the formation of the rock, and therefore how old it is. For ex-ample, after one half-life the ratio is 1:1, after two half-lives 1:3, and so on.

Even after geologists arrived at the presently accepted age of the earth of 4.5 billion years, demonstrating that Kelvin was off by a factor of 225, geologists continued to assume that Kelvin's error came from radioactive heat.[48] After all, it was the solution proposed by Ernest Rutherford, a Nobel Prize–winning physicist who would eventually be elevated himself to the rank of Baron Rutherford of Nelson. It was argument from authority all over again.

The actual flaw in Kelvin's model had nothing to do with unknown heat from radioactivity and everything to do with Kelvin's assumptions about the earth. The problem comes back to that contradictory-sounding term *fluid solid*. Kelvin assumed that the earth was a rigid solid, and that heat moved through it slowly. Put a poker into the fireplace, and the far end will heat up quickly, but it would take a long time for that heat to move up the poker and burn your hand. Before that happens, there will be a thermal gradient in the poker with one hot end, one cool end, and a transitional area in between. This slow motion of heat through a solid is called *conduction*.

Put a cauldron of water on that same fire, and something very different happens. Since the water in the cauldron is fluid, warm water at the bottom of the cauldron will rise, and cold water at the top will sink. It is the same pattern of convection that happens in the outer core, and it results in a much more even distribution of heat. There is little to no temperature difference between the top of the cauldron and the bottom.

Since the mantle is a fluid solid, and the core is incredibly hot, the mantle moves heat by convection. The top of the mantle is, therefore, just about as hot as the bottom of the mantle. When Kelvin was measuring the change in temperature with depth in the crust, the steep gradient he was measuring was not the result of a young earth, but of an earth where heat is distributed very evenly throughout its interior due to the convection of the mantle. (See figure 5.3.)

Unlike radioactive heating, the fluidity of the mantle was a well-established fact by Kelvin's day (see Pratt and Airy in the previous chapter). So the possibility of mantle convection should absolutely have occurred to a thermodynamicist like Kelvin. As a matter of fact, it did occur to Kelvin's lab assistant. In 1895, Kelvin's own assistant, John Perry, published two letters in the journal *Nature* pointing out that Kelvin's age of the earth would be too low if heat were moving more efficiently inside the earth than Kelvin thought, especially if the earth's interior was fluid.[49] Perry had the advantage of being right but the much greater disadvantage of being unknown. His conclusions were largely unacknowledged in his lifetime.

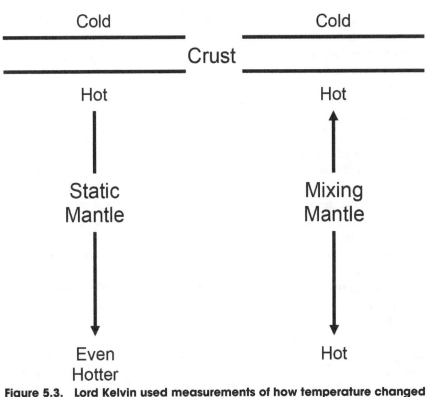

Figure 5.3.   Lord Kelvin used measurements of how temperature changed with depth in the crust to estimate how heat was distributed inside the earth. He assumed that the differences in the crust continued all the way down through the mantle (left). This led him to believe that the interior of the earth was incredibly hot and could not have been cooling very long. In actuality, because the mantle convects and mixes itself, heat distribution in the mantle is somewhat uniform (right). Kelvin's method for dating the earth might have worked, but it was based on flawed assumptions.

Young-Earth Creationism

The argument from authority fallacy occurs when you assume an argument is correct based on the source of the argument instead of its content. It led Ussher to use the Bible as a source of geological information. It resulted in Kelvin's age of the earth gaining acceptance despite the protests of geologists. It caused geologists to blame Kelvin's error on radioactivity for decades. And it continues to influence the debate over the earth's age today as it drives the young-earth creationism movement.

Young-earth creationism is exactly what it sounds like. It is the belief that the earth and all of the living things on it were created in their present forms within the past few thousand years. According to the National Center for Science Education, it is an opinion shared by about 40 percent of Americans.[50] For many of its adherents, young-earth creationism also assumes that you can learn most of what you need to know about earth history from the Bible. This assumption is front and center in the name of one of the largest young-earth creationist organizations in America today—Answers in Genesis. It is the cosmology of Sir Thomas Browne and James Ussher persisting in the present day. It also has an extensive infrastructure supporting it as a movement.

Answers in Genesis publishes the *Answers Research Journal* (ARJ), which is described on its website as a technical journal for professionals that publishes peer-reviewed articles relevant to a young-earth creationism perspective.[51] Its journal description is not all that different than the descriptions found on the websites for prestigious scientific journals like *Science* or *Nature*.[52] Most of the research agenda for young-earth creationists is intended to undermine the idea that the earth is old. They primarily publish attempts to disprove radioactive dating or to demonstrate that the entire rock record could have resulted from a single global flood. The fact that they need their own journal for these articles is a bit telling. Actual empirical evidence for either of these claims would radically overturn our notions of how nature works and would be accepted happily in almost any scientific journal. After all, "radioactivity doesn't work quite the way we thought it did and that has ramifications for our understanding of the age of the earth," is almost precisely the claim that earned Ernest Rutherford his renown and eventually a Nobel Prize.

Rutherford's work is a good point of comparison for the research done by young-earth creationists in that it actually did overturn conventional wisdom on the age of the earth. However, Rutherford was able to do this because his research was data-driven. The ratios of isotope to daughter product that he found in rocks led him to the conclusion that Kelvin was incorrect in his assessment of the maximum possible age of the earth. Rutherford could just as easily have dismissed his data because it disagreed with such an authoritative opinion. That wouldn't have been good science, but is exactly the strategy that young-earth creationists follow.

In the 1650s, James Ussher used the best data available to him to draw the conclusion that the earth was thousands of years old. In the early 1900s, Ernest Rutherford used the best data available to him to come to the conclusion that the earth was millions or even billions of years old. Young-earth creationism commits the argument from ignorance fallacy on a grand scale. It ignores all of our learning about the earth since the days of Ussher in order to reach his same conclusion. Young-earth creationists will never find data to contradict their assumptions because they are not looking for it. If they were, they wouldn't need to look much further than the nearest geology book.

CHAPTER 6

The Earth Has a History

If you had access to a time machine, you could travel back as much as 4.5 billion years into earth's past, but you probably wouldn't want to. At the time, the earth was a fiery ball of molten rock. You wouldn't last much longer on the earth as it existed 2 billion years ago either. The amount of oxygen in the atmosphere back then was only about 1 percent of current levels. You could spend some time 1 billion years ago if you packed a lunch. The lack of any plants or animals on earth at the time would make foraging for local food a challenge. Even if you only traveled back to 45 million years ago—just 1 percent of earth history—you would have trouble recognizing your surroundings. South America was an island, Europe was mostly underwater, and the Himalayas were barely foothills. Don't bother remarking about how different things were, though. There were no humans, and barely any other primates for that matter, to make conversation with.

The earth that we see around us today is not just old: it has a history. Continents have moved, life has evolved, and climatic patterns have shifted. Future chapters will discuss the mechanisms of those changes, but before you can talk about how and why things have changed, you need to realize that those changes have happened in the first place. While that might seem obvious to modern sensibilities, it is a relatively new realization.

In the industrial age, we have become used to constant societal change. A person born at the dawn of the twentieth century would have been born in a world of carriages and steamships and died in a world of airplanes and spaceships. But if you were born at the dawn of the seventeenth century, the world you died in would look almost identical to the world you were born in. It was actually a big leap for people to realize that the past was ever any different than the present. To make that realization about nature was particularly hard.

Our best evidence that the earth has changed over time comes from the long record of fossil organisms left behind in rock. Fossils let us know not only that life has changed but also that the physical conditions on earth have changed. Every species today has a particular habitat that it lives in, as did every species in the past. As an example, corals tend to live in tropical oceans. So if you find coral fossils in parts of Canada (and you do), that tells you that in the past, Canada was much warmer—and underwater.

Today, geologists use the unique sequence of fossils found on earth to keep track of the long expanse of geologic time and how the earth has changed over that time. But before they could do that, they needed to make several realizations. They needed to realize that fossils were the remains of living organisms, that most of them have gone extinct, and that you can put fossils into groups that lived at the same time.

Understanding What Fossils Are

Humans have been exploring the earth for hundreds of thousands of years, so many "firsts" in geology can't really be documented. Just like nobody knows who the first person was to notice that the earth is round, we really can't say who first discovered any particular mountain or any particular rock. That also goes for fossils. Nobody knows who the first person to find a fossil was, but we do know the first person to figure out what fossils are.

In 1666, Archduke Ferdinando II of Tuscany had just received a new personal physician—a Danish doctor named Niels Steensen (better known today by his Latinized name, Nicolas Steno). Steno had made a name for himself throughout northern Europe as an anatomist. He had already demonstrated that a cow's heart is made of ordinary muscle, and he had discovered a new salivary duct.[1] Steno had also made a reputation for himself as a freethinker. Steno was willing to question conventional wisdom and authority, having most recently questioned Descartes's assertion that tears originated in the brain.[2]

His new position in Tuscany permitted Steno a unique opportunity. In the fall of 1666, a group of Tuscan fishermen caught a monstrously large shark, and Ferdinando wanted his best anatomist to dissect its head for the public.[3] The dissection was an event. Steno did his work in front of a packed house. It was also an important event in the history of geology. During the dissection, Steno got a good look into the shark's mouth and realized that the teeth looked familiar.

Prior to 1666, people had been finding fossilized shark teeth for ages, but nobody was really sure what they were. This uncertainty is reflected in the choice of the word *fossil* to describe them. The word doesn't mean anything related to living things or petrification, or anything else you might assume. It comes from a Latin phrase meaning *dug up* and was originally used to refer to anything found in the ground, whether it was an interesting mineral specimen, an archaeological artifact, or what we would actually recognize as a fossil today—the remains of a once-living creature.[4]

In the first century CE, Pliny the Elder caused a sensation when he questioned the conventional wisdom that fossil shark teeth fell from the sky during a waning moon.[5] For most of the Middle Ages and Renaissance, fossilized shark teeth were called *glossopetrae* (tongue-stones), because they looked like tongues. Some people thought shark teeth were dragon tongues, while others suggested they were the tongues of snakes, dragons, or woodpeckers.[6] Leonardo da Vinci, who had more than a passing interest in fossils, included "describe the tongue of the woodpecker" on one of the to-do lists in his journals,[7] but it's hard to know if this was part of an investigation into glossopetrae or just Leonardo being Leonardo.

One of the most common interpretations of fossils during the Renaissance goes way back to the days of Plato and is somewhat hard for modern brains to wrap themselves around. According to Plato, all of nature is simply an imperfect manifestation of an ideal plane that exists outside of reality. All the different realms of nature do their best to show us the ideal forms that exist in this plane in their own way. This explains why you see repeated shapes across nature. Stars, star lilies, and starfish share their shape because they are the sky, the field, and the sea depicting the ideal form of the star. If you then find a fossilized starfish, you are not looking at the remains of a dead creature. You are looking at the rock world also showing us the same form.[8] Under this model, a fossilized starfish has no more relation to a living starfish than either has to a star lily.

In 1667, Steno published the results of his dissection of the shark's head in the imaginatively titled *The Head of a Shark Dissected*. In it, he argued that glossopetrae didn't just look like shark teeth: they were shark teeth that had been chemically altered and somehow preserved in rock.[9] It was one of the first claims that fossils were the remains of dead organisms.[10] It also got Steno puzzling over a question: How does one solid object get embedded inside another?

Steno's curiosity about the relationship of solid objects to one another resulted in the 1669 publication of the heftily titled *Prodromus to a Dissertation Concerning a Solid Body Enclosed by Process of Nature within a Solid* (typically just called *The Prodromus*). In *The Prodromus*, Steno spends considerable time speculating on the nature of rock outcrops and derives three laws for determining how these outcrops have developed over time. They are the basis for historical geology and still appear in geology textbooks as "Steno's Laws." They are also somewhat intuitive and derived from common knowledge.

If you have ever been to a beach or a desert, you may have noticed that the sand there tends to be spread out on the ground, and not stacked in vertical columns. That is Steno's first law. The Law of Original Horizontality and Lateral

Continuity says that sedimentary rock layers (those formed from sand and mud), start out horizontally.

If you have ever looked for something on a messy desk, you know Steno's second law. The longer it has been sitting on the desk, the deeper down in the pile it probably is. That is the Law of Superposition. Whether you're talking about books, pancakes, or rock beds, in any pile, the things at the bottom of the stack were there first.

If you have ever stapled sheets of paper together, you know that you need to stack and straighten the pages before you put the staple in. That's Steno's third law. The Law of Cross-cutting Relationships says that whenever one object cuts through another, the object being cut was there first. In other words, you can't cut into an object unless the object already exists.

Steno's laws are nicely encapsulated in the process of making a birthday cake. When you pour the batter into the pans, it spreads out horizontally under its own weight (originally horizontality). After baking the cake, you put down the bottom layer, a layer of frosting, the second layer, and then the final layer of frosting (superposition). Then, only after the cake itself is complete, you can put the candles in (cross-cutting relationships).

Prodromus literally means forerunner. Steno called his work *The Prodromus to a Dissertation* . . . because he planned to follow it up with a more detailed work. But he never did. Some authors have suggested it was because Steno understood, but couldn't accept, the ramifications of his own work.[11] By providing guidelines for reconstructing the history of a rock outcrop, Steno essentially began the field of historical geology. However, his work strongly suggested that the earth was old, and had formed over time. Coming to this conclusion would then require Steno to interpret some parts of Genesis as metaphorical, and we all know how well interpreting the Bible turned out for Galileo. Steno turned his back on the study of the earth and shifted his studies from anatomy and geology to theology. He converted from Lutheranism to Catholicism and entered the priesthood, quickly rising through the ranks. By the time he died, at the age of forty-eight, he had recently stepped down as the bishop of Titiopolis.[12]

If Steno really did abandon his geological pursuits because he thought that they would lead to heresy, then he fell prey to a slippery slope fallacy, which occurs when you reject taking a small step, either in action or in logic, for fear that it will lead to larger and potentially catastrophic steps in the future.[13] A common contemporary example of a slippery slope fallacy is rejecting the idea that cannabis may have legitimate medical uses out of fear that permitting medical use of marijuana may lead to legalizing harder drugs for recreational use.

While he might not have convinced himself, Steno did convince most geologists that the earth has a readable past and that fossils are a part of that past. But accepting that fossils are the remains of once-living organisms brings about new questions. Finding fossilized shark teeth is one thing. We've seen sharks and know what they are. But what do you do when you find fossils that don't look like any living animal?

The Discovery That Species Go Extinct

Sadly, no living person has ever watched a wooly mammoth walk across a meadow. Some enthusiastic geneticists are working to fix that fact,[14] but at the moment, mammoths are extinct. Velociraptors, Neanderthals, and millions of other species are also extinct. By some estimates, more than 99 percent of species that have ever lived are now extinct.[15] It's not necessarily a bad thing. The world changes, and life changes with it. But the fact of extinction wasn't always so obvious. In fact, when it was first proposed, the idea that an entire species could just vanish from the face of the earth was controversial—even revolutionary. But then again, that idea originated in revolutionary times.

If you had been born in France in late August of 1769, then by the time you reached the age of twenty-six, you would have lived under a feudal monarchy, a constitutional monarchy, a republic, and an authoritarian directory. You would have seen divinely anointed kings beheaded and lived through a Reign of Terror. You might not even be sure exactly how old you were since your birthday would now fall in the month of Fructidor, and suddenly people were declaring that it was the year 3. In short, you might yearn for stability, but experiences would cause you to suspect that you live in a world of repeated catastrophic change.

Georges Cuvier was born in France in late August of 1769; by the time he was twenty-six, he had developed a reputation as one of the greatest young anatomists in France.[16] As an anatomist, Cuvier was committed to the idea that each species on earth was stable, unchanging, and could be objectively described. However, his research was beginning to lead him to the conclusion that our modern world may have been shaped by a series of catastrophic upheavals.

Cuvier's nascent reputation, along with his position at Paris's National Museum of Natural History, gave him the credentials and the access to address a question that was vexing anatomists of the time: How many species of elephants are there? This was nowhere near as trivial a question as it might seem today because, for anatomists in the 1700s, there were no trivial questions. Nature was God's holy creation and was designed in accordance with His divine plan. Understanding nature was seen as a pathway to understanding that plan. If the world had two species of elephants, that was because God had reasons for there to be two species of elephants—not one, not three.

Unfortunately, determining how many species of elephants lived on earth was far more challenging in the 1790s than it might seem. At the time, species were determined by physical similarity. If two things met some minimum criterion of difference, then they were different species. This definition required trained anatomists to make careful examinations of multiple specimens and render their expert opinions. In the 1790s, however, most of these experts lived in Western Europe, a region notoriously devoid of elephants, so opportunities for making the necessary comparisons between specimens were few and far between.

Because opportunities for directly comparing an African elephant to an Indian elephant were scarce, biologists prior to Cuvier had used geography and behavior as a proxy for anatomy. They assumed that since the two populations of elephants lived so far apart from one another, and behaved very differently, they must be different species. One group of elephants lived in Africa, one in India. There were no elephants in between, so obviously, these were distinct populations and always had been. Modern biology supports this conclusion with evidence from genetics and reproductive behavior.

Beginning in the late 1700s, new fossil finds were beginning to undermine this conventional wisdom on elephants. Specifically, people were beginning to find the petrified bones of elephants in places where there were no living elephants, like Russia and New York. Scientists began to question whether the living elephants in India and Africa might be isolated surviving populations of a single species that had once lived all over the world.

Thanks to the plunders of France's recent military campaigns against its neighbors, the National Museum of Natural History had an impressive array of modern and fossil mammal specimens in its collections. In 1795, Cuvier went to work comparing the skulls of African elephants with Indian elephants. According to a speech he presented to France's National Institute in 1796, the investigation was a brief one. Cuvier claimed that he could tell "at a glance" that these were different species based on the characteristics of their molars.[17] What made his speech remarkable was when he turned his attentions to Siberian elephants (what today we would call mammoths). Cuvier argued that they represented a third species of elephant, as different from the other two species (in his words) as a jackal differs from a dog.[18] The only truly puzzling issue was what happened to this third species of elephant. Why did we find their bones, but no living animals?

Cuvier considered three possible explanations for the lack of living elephants in Siberia. He quickly dismissed the absolutely wonderful idea that since we only find the bones of Siberian elephants in the ground, then they must be burrowing elephants that we never see because they live underground. He also dismissed the idea that the Siberian elephant had somehow transformed into the modern African or Indian elephant. Cuvier was writing almost fifteen years before Charles Darwin's birth, but he did have some colleagues in Paris who were toying with the idea of species transformation. As an anatomist committed to the stability of species, however, Cuvier hated this idea. The only explanation that Cuvier could accept for the absence of living Siberian elephants was that there aren't any anymore. The entire species must be extinct.[19]

Claiming that an entire species had disappeared from the globe was unprecedented in 1796. It was a particularly bold pronouncement for a twenty-six-year-old whose life's travels had never taken him more than a few hundred miles from Paris. The map of the world, as it was known to Europeans in Cuvier's time, had a lot of blank spaces on it. There were a lot of places that mammoths could still be hiding. Cuvier, however, went beyond simply declaring the mammoth to be extinct. He declared it a remnant of an entire pre-human world that had been wiped out in a sudden cataclysm. He even suggested that the history of the world was a series of violent cataclysms, each wiping out the world before it and ushering in a new and improved version of nature.[20] It was a vision of natural history as tempestuous as the history that Cuvier himself had lived through.

Many contemporary scientists doubted Cuvier's conclusions. In fact, just one year after Cuvier declared the mammoth to be extinct, an American paleontologist would present a paper on the fossilized remains of a ground sloth arguing that since the bones were found in Virginia, then the living creature must be somewhere in the same area.[21] This paleontologist didn't have the same reputation as an anatomist that Cuvier did, but he was still relatively well known. At the time, this particular paleontologist, Thomas Jefferson, was the vice president of the United States.

Resistance to Extinction and Geological Change

There are very few scientists who have ever appeared on U.S. currency. In fact, among the faces gracing the money currently in circulation, you will find exactly two. The hundred-dollar bill features the serene smile of Benjamin Franklin, who advanced our understanding of electricity and who was a prolific inventor. On both the two-dollar bill and the nickel, one finds the agronomist, naturalist, inventor, paleontologist, and (of course) vice president, president, and statesman Thomas Jefferson. To say that Jefferson lived through revolutionary times is almost unnecessary when discussing his political career. It is, as Jefferson himself might have put it, self-evident. However, there were also revolutions in our understanding of nature during his lifetime that were just as sweeping. Jefferson was born into a world that everyone knew was about ten thousand years old and had always existed in its present state, and he died in a world that was unknowably old and constantly changing. Jefferson kept up with many of the changing scientific attitudes of his time, and sometimes that got him into trouble in his political career.

In the run-up to the presidential election of 1804, in which Jefferson would defend his presidency against the campaign of Charles Pinckney, an anonymous pamphlet began circulating in New York attacking Jefferson for his heretical scientific views. The pamphlet's title summed up its conclusions: *Observations upon Certain Passages in Mr. Jefferson's Notes on Virginia, which Appear to Have a Tendency to Subvert Religion, and Establish a False Philosophy* (hereafter, just *Observations*).[22] Its author claimed that many of the observations made by Jefferson in his 1781 book *Notes on the State of Virginia* implied that the earth was old and that it had developed over time, and that if Jefferson truly believed these things, he was unfit to continue serving as the nation's president.[23]

The anonymous author of *Observations* cites two specific bits of Jefferson's writing that imply an old and changing earth. In a discussion of the indigenous languages of America, Jefferson points out that dialects typically take centuries or even millennia to differentiate themselves from one another and develop into fully formed languages. The existence of so many different languages among the indigenous people of America implies that the process of linguistic separation must have played out dozens of times over.[24] *Observations* points out that this would require the earth to be many times older than the few thousands of years that the Bible implies.[25]

Jefferson compounded his alleged sins in his description of the Shenandoah River Valley, in which he writes that "the first glance of this scene hurries our senses into the opinion, that this earth has been created in time, that the mountains formed first, that the rivers began to flow afterwards."[26] This, according to *Observations*, is just flat-out heresy. It is in direct opposition to the revealed word of Genesis, which clearly states that the waters were gathered together before the land was exposed.[27]

The author of *Observations* concludes by revealing the origin of Jefferson's heretical notions. Jefferson, they say, has been unduly influenced by the French. Jefferson was a well-known Francophile, and several contemporary French philosophers, including those who had influenced Cuvier's view of nature, were beginning to develop theories about the earth's formation that implied a longer history than Scripture strictly allowed. *Observations* sums it up thusly:

> It must be confessed that the author of the Notes on Virginia is one of the most confused and unintelligible writers that ever the world produced. And all good Christians should ardently hope that he would prefer the imputation of being a bad composer, to the suspicion of being a humble follower of modern French philosophers. Wretched, indeed, is our country, if she is going to be enlightened by these philosophers. . . . They build up theories of shells and bones

and straws. And for what? Is it to render more stable the uncertain condition of man? Is it to alleviate one of the miseries which afflict his nature? No; it is to banish civilization from the earth.[28]

The chain of logic is unbreakable. If you believe that rivers erode mountain valleys, then you are in direct violation of God's revealed word, which means you must be an immoral follower of French philosophy, and therefore you are out to destroy civilization. It would be laughable logic if its demonizing didn't ring so familiar to modern ears. It's also another example of a slippery slope fallacy. The logic of *Observations* is the same logic that ended Steno's geological investigations, just taken a few steps further.

If you've never heard of President Charles Pinckney, that's in part because *Observations* was unsuccessful in derailing Jefferson's quest for a second term. *Observations* was ineffective as a political argument, but it does show the passions that could be incensed by geological arguments at the beginning of the nineteenth century. It is also noteworthy for another reason. Eventually, Americans learned the identity of its author: Clement Clarke Moore, who is better known today for a second publication of his (also initially published anonymously), containing one of the most famous opening lines in the history of American poetry, "'Twas the night before Christmas, and all through the house."[29] Twenty years after trying to end the political career of Thomas Jefferson, Moore wrote "A Visit from St. Nicholas." The man who created our modern notion of Santa Claus was worried that Jefferson's writing might someday change the way we observe Christian tradition.

While Thomas Jefferson was willing to see the landscape of Virginia as a work in progress, he did not extend that belief to its wildlife. Jefferson's admiration for French natural philosophy stopped short of embracing Cuvier's extinction hypothesis or any of Cuvier's contemporaries' similar beliefs in previous iterations of nature. Jefferson believed that the earth had been different in the past but rejected the idea of extinction. He held to a conception of nature called the "Great Chain of Being," which was popular in the 1700s and which saw all of nature and natural history as an unbroken continuum.[30] For Jefferson, change in nature was possible only if it was slow and gradual. The idea that there had been previous versions of nature, separated from today by violent cataclysms, would have been anathema to him.

Jefferson had one other major area of disagreement with French natural philosophy. A commonly held belief in Europe was that American plants and animals existed in a state of degeneracy relative to their European counterparts. That is to say, creatures in the new world were smaller, weaker, and generally less impressive than those in Europe. This theory was popularized by the most revered French naturalist of the day, the Comte du Buffon, and it offended

Jefferson's sense of patriotism.[31] During the years that Jefferson lived in Paris, he frequently debated Buffon on this exact point.[32]

In 1796, Jefferson received a package from Colonel John Stewart containing some fossilized leg, arm, and claw bones that had been excavated by miners in what is today West Virginia. Today we would recognize these bones as belonging to a ground sloth, a ten-foot, two-thousand-pound monstrous herbivore that used to roam the Americas. Jefferson, however, described them as the remains of an enormous lion, which he called *Megalonyx* (or "giant claw").[33] The description of *Megalonyx* that Jefferson submitted to the American Philosophical Society contains two telling details. First, he is delighted to be able to show Buffon a lion larger than any ever known from Europe or Africa. Jefferson was so eager to refute Buffon's ideas that the word *Buffon* actually appears more frequently in his description of the specimen than either the word *lion* or even *Megalonyx*.[34] Second, he is excited for the inevitable discovery of living *Megalonyx* somewhere in North America.[35]

By the early 1800s, geologists in Europe had mostly accepted the factuality of extinction, and (as we shall soon see) were beginning to use that fact as the basis for subsequent discoveries. However, Jefferson was a holdout for a long time. He firmly believed that if a species existed in the past, it must still exist, if undiscovered, today. The letter of instruction that Jefferson gave to Lewis and Clark made it clear that he expected them to bring back stories of living mammoths and *Megalonyx* that were thriving in the Pacific Northwest.[36] He was disappointed, to say the least.

For most of his life, even after most scientists had accepted it as a fact of nature, Thomas Jefferson rejected the idea of extinction. The only evidence that Jefferson may have ever come around on the idea of extinction comes from the very end of his life. In 1823, at the age of eighty, Jefferson wrote a letter to John Adams meditating on the inevitability of death, in which he cited the extinction of stars and of species as proof that everything eventually ends.[37]

The Formation of the Geologic Time Scale and the Importance of Passionate Amateurs

In the first half of the nineteenth century, British science got organized. Almost every major scientific society in London formed in this fifty-year span. For example, the Botanical Society, the British Science Association, the London Electrical Society, the Royal Agricultural Society, the Microscopical Society of London, the Entomological Society, the Royal Geographical Society, and the Royal Statistical Society were all formed just within the decade of the 1830s. This frenzy of society-building didn't come out of nowhere. This was at the same time that a wave of democratizing revolutions was sweeping America and Europe. The establishment of all of these societies in science may have been a reaction to this trend toward democracy in society—an attempt to keep science the purview of the privileged few.

One of the first scientific societies of this era was the Geological Society of London (GSL), founded in 1807. Like many other scientific societies of the time, the GSL was founded by an elite group of upper-class gentlemen. Its thirteen founders included a member of parliament, the secretary of the Royal Society, three doctors, several chemists, and a French count.[38] Unlike many of the other scientific societies of the time, however, the GSL at least professed aspirations of being inclusive of people from all social standings. In fact, they saw getting the perspectives of a broad swath of society as crucial to the business of understanding the earth. One of the GSL's first publications was a pamphlet for new members titled *Geological Inquiries* that included the following passage in its introduction:

> To reduce Geology to a system demands a total devotion of time, and an acquaintance with almost every branch of experimental and general Science, and can be performed only by Philosophers; but the facts necessary to this great end, may be collected without much labour, and by persons attached to various pursuits and occupations; the principal requisites being minute observation and faithful record. The Miner, the Quarrier, the Surveyor, the Engineer, the Collier, the Iron Master, and even the Traveller in search of general information, have all opportunities of making Geological observations; and whether these relate to the metallic productions, the rocks, the strata, the coal of any district; or to the appearances and forms of mountains, the direction of rivers, and the nature of lakes and waters, they are worthy of being noticed.[39]

The founders of the GSL knew that essential geological observations were being made daily throughout the country by working-class Britons, and welcomed those observations into the body of geological knowledge.

Exactly how welcoming they were of the people making those observations is a bit more debatable. While there were no overt class restrictions in society membership, the monthly meetings were dinner meetings, and the dinner cost 15 shillings.[40] While this was hardly a fortune (about $20 in today's money), it would have been a prohibitive price for your average miner, quarrier, or collier of the day. The egalitarian trappings of the GSL are also called into question by the way they treated one particular laborer—a canal engineer named William Smith.

William Smith was born in Oxfordshire in 1769. By his own account, his formal education was meager and often interrupted of his own accord. He would

frequently leave the local school to wander the countryside in pursuit of interesting rocks and fossils.[41] While his formal education may have been wanting,
Smith was renowned for his memory. Even in old age, Smith would often tell
stories of his youth in astounding detail. And he never forgot where he found a
particular fossil.[42]

In 1787, Smith met Edward Webb, a surveyor working in the area around
Littleton. Although Smith had no formal training in surveying, his enthusiasm,
intellect, and interest in geology impressed Webb.[43] Webb, like Smith, had little
formal education, and he was so impressed by Smith that he took Smith on as his
assistant on the same day that they met.[44] In 1791, Webb sent Smith to survey
an estate in Somerset that had recently been inherited, along with several coal
mines in the area, by a new owner. Smith's work was of such high quality that
the new estate owner gave him additional work surveying a canal for some of
her newly acquired coal mines.[45] It was in this canal that William Smith would
eventually discover the basis for geologic time.

Imagine that you have been given the loose pages of a book. Each page
describes one specific scene in detail. Put all those pages together and it tells a
fascinating story. Unfortunately, you have been given those pages in random
order. That was essentially the state of geology at the beginning of the nineteenth
century. Geologists could look at individual rocks, fossils, and landscapes and
use them to reconstruct a snapshot of earth history, but they had no reliable way
to put those vignettes into chronological order to tell a coherent story. Thanks to
Steno, they had some tools that they could use to put events in order across small
geographic areas, but telling the story of the earth required being able to put all
of those local histories together in the right order—a process called correlation.
Prior to William Smith, geologists had no way to correlate rocks worldwide.

While working in the canals and coal pits of Somerset, literally immersed
in the layers of the earth, Smith made a simple discovery with profound consequences for the ability to correlate rock. Specifically, while surveying, Smith
discovered that each of the different layers of rock he was working through had
its own unique group of fossils. Smith made extensive collections of the fossils
that he found in the areas where he worked. He organized these collections based
on the precise layers of rock that he found them in and displayed the collection
in his home. Using Smith's collection as a reference, miners were able to identify
the layers of rock they were working in far more precisely than ever before.

Correlating rock layers based on the fossils they contain is called *biostratigraphy*, and it is particularly useful because fossils tend to have very broad geographic ranges. Local geology can tell you if the rocks in one part of Somerset
are older or younger than rocks in another part of Somerset. Biostratigraphy can
tell you if the rocks in Somerset are older than the rocks in Wales, Scotland,
or even China. Smith had discovered the key to correlating rocks across great

distances. Geologists quickly realized they could use that key to put the pages of the geologic story in proper order.

Geologists around the world used the new tool of biostratigraphy to commence work on a global geological time scale. That time scale, which is still in use today, was based entirely on fossils. The unusual names that populate the scale (e.g., the Jurassic, the Mississippian, the Permian) mostly come from the places where fossils of that age were first discovered (in this case, the Jura Mountains of France, Mississippi, and the Perm region of Russia). An unusually large number of those names (including the Devonian, the Oxfordian, the Bathonian, the Kimmeridgian, the Cambrian, and many others) come from places within a hundred miles of Smith's old shop in Bath and are a testament to his work.

William Smith spent years working to put all of his information about fossil occurrences and rock correlation into map form. A single map showing the order of every rock layer in England would be invaluable to miners, surveyors, or almost anyone else whose work involved finding specific rocks or rock types in the earth. Many people knew about his work, but Smith could not find a publisher until 1812.[46] The map was published in 1815 and is considered one of the most important contributions ever made to geology as a science. Unfortunately, by this time, Smith was deeply in debt. He had dedicated so much time and effort to working on his map that his surveying work had fallen by the wayside for years. He had to sell his fossil collections and his personal library, and in 1819 he was sentenced to ten weeks in debtors' prison.[47]

The founding document of the GSL recognizes the importance of practical workers in the gathering of geological data. William Smith was a practical worker who gathered data that would revolutionize geology and make it possible for geologists to work on a truly global scale. Yet the relationship between Smith and the GSL was, for lack of a better word, a rocky one. Historians debate the source of the friction, but it appears to have been a personal dispute between Smith and the GSL's first president, George Bellas Greenough, who was working on his own geological map at the same time that Smith was making his.[48] Because of tension between Smith and Greenough, the GSL didn't acknowledge Smith's contribution to the construction of the geologic time scale until 1831. But in that year, the new president of the GSL, Adam Sedgwick, gave Smith the society's highest honor and described him as "the Father of English Geology."

Is Change Slow or Sudden?

In his 1748 work *An Enquiry Concerning Human Understanding*, Scottish philosopher David Hume asserts that most of the cause-and-effect relationships that we see in the world are, at least in part, figments of our imagination.[49] He famously uses the example of one billiard ball striking another. According to Hume, when the second ball begins to move, all we can observe about that motion is that it happened *after* the two balls collided—not *because* they collided. Any attempt to invoke a cause-and-effect relationship requires us to invent something that we cannot directly observe, such as momentum, force, or kinetic energy. He claims that the reason that we invoke cause-and-effect relationships is experience. Since we always see the second ball move after the first ball hits it, we assume that one had something to do with the other. Hume writes:

> For all inferences from experience suppose, as their foundation, that the future will resemble the past, and that similar powers will be conjoined with similar sensible qualities. If there be any suspicion that the course of nature may change, and that the past may be no rule for the future. all experience becomes useless, and can give rise to no inference or conclusion.[50]

Hume had no way of knowing this, but his insight that we can only function in this world by assuming that the future will be like the past would form the basis for scientific geology.

Geologists use Hume's axiom in reverse to create their motto, "The present is the key to the past." This motto, while ubiquitous in geology textbooks, appears to be the scientific equivalent of "Play it again, Sam," or "Luke, I am your father," in that everybody quotes it, but nobody actually said it. It is most often attributed to the Scottish geologist Charles Lyell, but he never appears to have used this pithy phrase. About the closest he ever came was in his 1863 book *The Antiquity of Man*, in which he wrote "The system of changes now in progress in the organic world would afford, when fully understood, a complete key to the interpretation of all the vicissitudes of the living creation in past ages."[51] While Lyell may never have actually written "The present is the key to the past," he certainly popularized the idea.

Lyell's best known and most significant work is the four-volume *Principles of Geology*. Its importance is perhaps best summarized by its subtitle: *An Inquiry How Far the Former Changes of the Earth's Surface Are Referable to Causes Now in Operation*.[52] Lyell was addressing a fundamental epistemological problem in geology: Geologists can't see things happening in the past. We can only see the consequences of those past actions in the present. In *Principles of Geology*, Lyell makes the argument that geologists can learn about the past by studying the present because all of the causes that have acted in the past are still acting in the present. For example, if you understand how streams move pebbles today, then you also know how streams have always moved pebbles. The world still works in exactly the same way it always has.[53]

Lyell's principle, that studying present causes can help scientists understand the earth's past, made a scientific geology possible, but it also placed an unnecessary hindrance on the scope of geological explanations. In *Principles of Geology*, Lyell asserts that the laws of nature are the same as they have been in the past, that natural causes are the same as they have been in the past, and that natural rates of change are the same as they have been in the past. Those first two assertions are important underpinnings for any scientific study of the past. But that third one isn't necessarily true.[54] In fact, it even seems unlikely. No river has ever flowed faster than the Amazon flows today? Mount Everest is the tallest mountain ever? No desert has ever expanded faster than the Sahara does today?

Lyell didn't include this third claim about the earth's past out of the blue. He was actually taking a position on a major geologic controversy of his day: Were the changes that occurred in the earth's past gradual or sudden?[55] By the 1830s, most geologists accepted that the earth had a history, but not everyone agreed about the tempo of that history. Were geological changes slow and steady or sudden and episodic? By including a preference for gradual explanations into *Principles of Geology*, Lyell was implying that they were the only truly scientific explanations in geology.[56] This is another example of the fallacy of persuasive definition, and it was remarkably effective. Lyell's model of geological change would eventually be given the name Uniformitarianism—a word that implied slow constant change as the only proper mode of explanation in geology. Geologists assumed that catastrophic events did not happen in the past—until they found evidence for a spectacular one.

Just outside of the village of Gubbio in the Apennine Mountains of Italy sits one of the most famous rock outcrops in all of geology. It is one of only a few dozen places on earth where rocks from the end of the Mesozoic Era (sometimes known as the Age of Dinosaurs) are exposed at the surface of the earth. In this rock outcrop is a thin layer of clay. The rocks below that clay layer formed while dinosaurs walked the earth. Those above it formed after their extinction.

In June of 1980, a paper appeared in the journal *Science* demonstrating that the clay layer in Gubbio was highly enriched with elements that don't occur naturally on the surface of the earth. It concluded that the only way to explain this enrichment was if a massive extraterrestrial object had struck the earth at the end of the Mesozoic.[57] It was the first good evidence that an asteroid impact killed the dinosaurs. This hypothesis has since been informally named "the Alvarez hypothesis," after the lead authors of that 1980 paper.

Over the next several years, sedimentary geologists would discover that this layer of meteoritic clay extended all around the world at the end of the Mesozoic. Mineralogists would find other minerals in the clay that had previously only been known from atomic bomb test sites and meteor impact craters.[58] Paleontologists would find evidence for a long shutdown of photosynthesis on earth.[59] Planetary geologists would find a 150-mile-wide crater dated to the end-Mesozoic age off the coast of the Yucatan Peninsula.[60] Nearly all of the different branches of geology came together and told one coherent story: one of catastrophic mass extinction. The extinction of the dinosaurs wasn't a slow gradual process, as Lyell would have insisted. From a geological perspective, it happened in the blink of an eye. Geological change was not limited to slow and steady after all. For 150 years after Lyell published *Principles of Geology*, geologists had rejected all catastrophic models for past events, until the evidence for a massive asteroid impact became nearly incontrovertible.

Lyell is still rightly considered one of the most important geologists in the history of the discipline. He is therefore also one of the most honored. He is buried in Westminster Abbey, is the namesake of a town in New Zealand and a valley in Yosemite, and—perhaps most ironically considering the consequences of the Alvarez hypothesis—has impact craters named for him on both the moon and Mars. His work cemented the connection between the present and the past that allows geology to proceed with a rigorous scientific method and allowed geologists to begin thinking about how changes have happened in the past. The processes that have caused geologic change are the subject for the remainder of this book. Thinking about these processes, however, requires thinking a little bit differently about the scientific method. Words we often associate with science—like *observation*, *experimentation*, and *repeatability*—need serious reconsideration when trying to create a science of the past.

Part II

CHANGE

CHAPTER 7

Our Methods Change

One of my high school teachers once told me that if I ever found myself in a situation while driving where a head-on collision was inevitable, I should speed up. That way, I would hit the other car harder than they hit me. This was, of course, terrible advice, and I'm happy to say that the teacher in question did not teach Drivers' Ed. Newton's third law of motion makes it clear that for every action, there is an equal and opposite reaction. The harder you hit something, the harder it hits you back. Knocking your head against a wall doesn't just hurt the wall. Newton's third law is a fundamental law of nature, but why is it true? Why can't you hit the other car harder than it hits you?

After he published *The Principia*, people asked Newton almost that exact question: Why are the laws of motion the way they are? His answer, perhaps surprisingly, was that it was none of his business. In an addendum to the second edition, Newton writes:

> I have not as yet been able to discover the reason for these properties of gravity from phenomena, and I do not feign hypotheses. For whatever is not deduced from the phenomena must be called a hypothesis; and hypotheses, whether metaphysical or physical, or based on occult qualities, or mechanical, have no place in experimental philosophy.[1]

If that attitude, that hypotheses have no place in science, sounds strange today, that's because it is fundamentally different from the way we do science today. Every grade school student learns "the scientific method." But, in actuality, what they are learning isn't *the* scientific method; rather, it is *a* scientific method—the one we happen to use now. There have been others in the past.

One of the most innocuous yet loaded sentences you can use in an academic setting is "that's not science." It's innocuous because for most things it's true.

Most fields of study and most topics of conversation have nothing to do with science. Yet this statement feels dismissive, obnoxiously so. Imagine the following conversation:

Person A: What do you do here at the college?

Person B: I'm in the literature program. I'm a Melville scholar.

Person A: That's not science.

Person A is stating a fact, and probably one that person B would agree with. And yet you can't help but feel that person B has been insulted and that person A is being argumentative.

All good definitions help you decide not only what a word refers to but also what it does *not* refer to. The definition of *science* is no different. Since the time of Thales, scientists and natural philosophers have struggled to determine what does and does not qualify as a proper method for explaining nature. As a result, definitions of *science* have changed repeatedly over time, sometimes to such a degree that they would exclude past scientists. Aristotle would undoubtedly have said, "That's not science," to Isaac Newton, and Isaac Newton would have been equally dismissive of Charles Darwin.

The most long-lived scientific method was developed by Aristotle in the third century BCE and was the final word on how to do science for nearly two thousand years. Aristotle's scientific method began from the assumption that true knowledge was universal. That is, the most important things were the things that are always true. For example, all birds have feathers. Once you have this universal knowledge, you can apply it to learn things about specific cases. You may have never heard of or seen a California Clapper Rail, as it is one of the rarest birds in America. But you still know something about it: it has feathers.

One interesting thing to note about Aristotle's universals is that you don't need to make observations to use them. In fact, it is almost impossible to observe a universal. Nobody has observed all the birds. Therefore, a statement like "all birds have feathers" could never come from observations. That's what makes Aristotle's method of science seem so strange to modern readers. It is a comprehensive system for understanding and organizing our world that has very little need for actually observing the world.

Science without need for observation seems like a contradiction today, yet it was the preferred method of doing science for nearly two thousand years. The awe in which people held the classical philosophy of Ancient Greece meant that Aristotle and his method enjoyed a combination of both the argument from authority and a second fallacy to maintain their primacy in science. The appeal to tradition fallacy is a form of intellectual inertia.[2] It is the argument that we should continue to do something in a certain way because it has always been done that way. If the Aristotelian method was good enough for classical Greek scholars, classical Islamic scholars, medieval theologians, and literal Renaissance men, then it should be good enough for us. Fortunately, though, by the 1600s, natural philosophers were finally beginning to consider an alternative.

When science moved away from Aristotle, the pendulum swung hard in the other direction. In 1620, Francis Bacon published his *Novum Organum*. It included what Bacon considered to be the proper rules for investigating nature. For him, pure observation was the key. The job of a scientist is to observe nature, and then report their observations. The last thing you would ever want to do is to interpret. According to Bacon, once you start interpreting you stop describing nature and begin imposing your own biases and opinions upon it.[3] Stick to the facts, and keep your opinions to yourself.

In the hagiography of British Science, Bacon explained the rules for doing proper science, and then Newton showed their power.[4] That's why Newton refused to answer questions about why the rules were what they were. He would not explain the reasons behind the laws of motion because any such explanation would cause Newton to go beyond what he had actually observed. He feigned no hypotheses. In Latin, that phrase translates as the delightful-sounding "*hypotheses non fingo.*" This motto would set the rules for what qualified as a scientific explanation in science for centuries.

When the Geological Society of London was established, one of its guiding principles would be that it was established for the dissemination of facts, not the discussion of theories. After all, if geology was to be a proper science, it needed to be based on good Baconian observations and Newtonian method, not on speculation and theory. In an 1861 letter to his friend, the economist Henry Fawcett, Charles Darwin comments on how such a strict adherence to Bacon's guidelines made geology not only difficult but boring as well. He writes:

> About 30 years ago there was much talk that geologists ought only to observe and not theorise; and I well remember someone saying, that at this rate a man might as well go into a gravel pit and count the pebbles and describe their colours. How odd it is that everyone should not see that all observation must be for or against some view, if it is to be of any service.[5]

If Darwin could express this sort of frustration at the Baconian method of simply reporting your observations, then at some point in the mid-nineteenth century, the pendulum of what qualified as science must have swung again. Pure observation, with no interpretation, was no longer considered the one true path of scientific investigation. The reason for this change is one of the most accomplished Victorian scientists that almost nobody has heard of, William Whewell.

The list of William Whewell's accomplishments is almost obnoxiously long and varied. He was a poet, an Anglican priest, and a professor of both mineralogy and philosophy at Cambridge. His published works include books on German church architecture, physics, astronomy, education, exobiology, oceanography, and economics, as well as an original translation of *Plato's Republic*.[6] Some sources describe him as second only to Shakespeare in contributing words to the English language. But one of the things that fascinated Whewell the most was science itself—its history and its philosophy.

In a pair of massive tomes titled *History of the Inductive Sciences, from the Earliest to the Present Times* (1837) and *The Philosophy of the Inductive Sciences, Founded upon Their History* (1840), Whewell argued that there were two ways to investigate nature. The first method he called "colligation of facts." This is the classic Baconian method of making a series of repeated observations and extrapolating from them a general law. It works wonderfully when you have the opportunity to make a series of repeated observations.

The second method Whewell called "consilience of inductions." Consilience occurs when multiple different lines of evidence all lead you back to the same conclusion. Whereas colligation asks the question, "What would explain all of my observations?" consilience might ask, "What observations would support my explanation?" According to Whewell, different disciplines are better served by different methods. Whereas a physicist might use colligation to study ions, another scientist might use consilience to study linguistics. Incidentally, *physicist, ion, scientist, consilience*, and *linguistics* are all examples of words that Whewell coined.

The most commonly used (and taught) scientific method today goes by the impressive name of the hypothetico-deductive method.[7] This method begins with a theory, usually based on some observations, and considers the ramifications of that theory. Essentially it asks, "If my explanation for what I'm seeing is true, then what else must be true, and what else must be false?" It then searches for those things that must be false. As the philosopher of science Karl Popper once put it, "Every good scientific theory is a prohibition: it forbids certain things to happen. The more a scientific theory forbids, the better it is."[8] Science proceeds this way because it is much easier to prove a theory false than to prove it true.

Depending on when and where you lived, you could spend an entire lifetime convinced that it always snows at Christmas. Suppose, however, you wanted to test the theory that it always snows at Christmas. No number of snowy Decembers would actually prove that *every* Christmas is snowy, but one balmy one would instantly disprove it. In this case, a single green Christmas is a much more important data point than a lifetime of white ones. Then, when that green Christmas came, you would be required to do something very hard. You would need to set aside your theory about Christmas always being white.

Whewell's concept of consilience, along with the hypothetico-deductive method, allows geology to ask questions about earthly processes that happen beyond our ability to directly observe them—either because they only occurred in the past, or because they happen deep in the earth, or because they happened too slowly to be observed in a human lifetime. Geologists supplement their direct observations with an understanding of the consequences of their theories. For example, we have (thankfully) not directly observed an extinction-sized meteorite hitting the earth within the span of human history. However, we can predict what evidence such an event would have left behind if it had happened in the past and look to see if that evidence exists.

What Whewell introduced to geology was an ability to ask questions about cause and effect in the past. He named this undertaking *palaetiology*, literally "the study of ancient cause," because it needed a name and making up words is what he did.[9] This opened up entirely new questions for geologists to study. The field was no longer limited to answering questions about how the earth is today. Instead, Whewell's work provided an intellectual framework for discovering how the earth might have been different in the past.

What geologists have discovered is that while many things about the earth have remained constant over time, including its shape, its size, its relative position in the universe, and all of the other topics of the previous chapters, quite a bit has in fact changed: the arrangement of continents, the creatures living on and around those continents, the environments in which those creatures lived, and so on. In the next several chapters, we will explore all of those changes and the controversies that have accompanied their discoveries.

CHAPTER 8

Life Changes

Many popular accounts of the life of Charles Darwin include a description of an *aha!* moment, in which Darwin suddenly has his great insight into the mechanism of evolutionary change. Tellingly, these accounts vary on what that moment was. Some claim it happened in the Galapagos Islands while he was studying the finches. Others claim it happened years later in a library where he was reading economic theory and realized the same principles applied to nature. In either case, it would be ironic, bordering on weird, if the theory of evolution suddenly sprang forth all at once in its current form in a single moment of inspired creation. In actuality, the development of natural selection as an evolutionary mechanism came from generations of work, including multiple generations within Darwin's own family. It also came about as a result of Darwin's lifetime of experiences.

Darwin faced two great challenges to his theory. When he first published *On the Origin of Species* in 1859, he was accused of a type of heresy—but not religious heresy. Early critiques of Darwin's work argued that it wasn't Baconian enough to qualify as good science because it did not rely on direct observation. At first, religious critiques of Darwin were much rarer than methodological critiques.[1] Large-scale religious opposition to Darwin came much later and was founded on fears that were not necessarily rooted in what Darwin actually wrote. The American anti-evolution movement began as a response to misreadings and misapplications of Darwin, and only later evolved to address the substance of Darwin's argument.

Darwinism Before Charles Darwin

Almost exactly one hundred years to the day before Charles Darwin left for his five-year voyage on the *HMS Beagle*, his grandfather Erasmus Darwin was born in Elston, England.[2] He was the seventh child and fourth son of Robert Darwin, a prominent lawyer from a distinguished family. Erasmus Darwin chose a different path from his father and, after receiving his medical degree at Edinburgh, settled into private practice as a physician in Lichfield. Erasmus Darwin built a practice of some renown over his fifty-year career in medicine, and at one point even turned down an invitation to become a personal physician to King George III.[3] However, what he was best known for in his day was his poetry.

Erasmus Darwin's poetry was moderately popular among the public, but not well-liked among his contemporary poets. Lord Byron called him "a mighty master of unmeaning rhyme."[4] Samuel Taylor Coleridge remarked (a little more vividly) in a letter to a friend, "I absolutely nauseate Darwin's poems."[5] Darwin's poems are often sickly sweet. Consider the following passage from his 1789 poem "The Loves of Flowers":

> With secret sighs the Virgin Lily droops,
> And jealous Cowslips hang their tawny cups.
> How the young Rose in beauty's damask pride
> Drinks the warm blushes of his bashful bride.[6]

It goes on like that for hundreds of lines, describing the romantic lives of the various plants in his garden.

If Erasmus Darwin had limited his poetry to botanical bodice-ripping, it's likely that he would be best remembered today as a prominent Georgian Era doctor and little more. But a few lines down, he provided a hint of his work to come:

> BOTANIC MUSE! who in this latter age
> Led by your airy hand the Swedish sage,
> Bad his keen eye your secret haunts explore
> On dewy dell, high wood, and winding shore.[7]

The "Swedish sage" in this stanza refers to Carl Linnaeus, the Swedish botanist who created the "Kingdom, Phylum, Class, Order, Family, Genus, Species" system of biological classification still in use today. Erasmus Darwin wasn't just using his poetry to describe nature. He was using it to comment on contemporary theories of nature and to advance a few of his own.

It may seem unusual today to write scientific theory and commentary in the form of romantic poetry, but Erasmus Darwin was unusual in a lot of ways. He was an abolitionist in the 1700s.[8] He believed strongly that women should be educated.[9] He enjoyed fine food to the point that he had a semi-circular hole cut out of his dining table so that he could get closer to his meals. He had at least fourteen children by at least three different women.[10] But perhaps most unusually of all, Erasmus Darwin was a transformationalist. To use a more modern phrasing, Erasmus Darwin believed in evolution before it was cool.

One very common misconception about evolution (or, as it was called in Erasmus Darwin's day, *transformationalism*) is that Charles Darwin invented the idea. The notion that life can change over time goes back at least to Ancient Greece. The pre-Socratic philosopher Anaximander, back in the 500s BCE, believed that since humans are born helpless, the first humans must have been born to some other species of animal. He also suspected that life began with fish and moved onto land later.[11] By the late 1700s and early 1800s, when Erasmus Darwin was writing, transformationalism was a fringe idea in biology, with most of the more prominent transformationalists living in France. This fact, of course, only made the idea even more suspect to British scientists.

Erasmus Darwin made no secret of his transformationalist proclivities. In *The Temple of Nature*, published shortly after his death, he wrote:

> Organic life beneath the shoreless waves
> Was born and nurs'd in ocean's pearly caves;
> First forms minute, unseen by spheric glass,
> Move on the mud, or pierce the watery mass;
> These, as successive generations bloom,
> New powers acquire and larger limbs assume;
> Whence countless groups of vegetation spring,
> And breathing realms of fin and feet and wing.[12]

It should be noted that this is absolutely not the evolution by means of natural selection that Erasmus's more famous grandson would eventually propose. Like many of his contemporaries, Erasmus Darwin believed in spontaneous generation; that is, he believed that simple life could just originate fully formed from lifeless matter. He also believed, as would the noted French transformationalist Jean-Baptiste Lamarck, that, over time, living things could acquire new characteristics and pass them on to future generations. Therefore, in Erasmus Darwin's model, complex organisms (those with "fin and feet and wing") were the descendants of some of the first creatures to precipitate out of seawater. Conversely, the simpler organisms that live today are the descendants of those "forms minute" that originated more recently.

There were few transformationalists in England before Erasmus Darwin published his poetry, and probably equally few after. His theories were seen as more entertaining than persuasive. The most common contemporary idea of how nature came about was detailed in an 1802 book by the Reverend William Paley titled *Natural Theology; or, Evidences of the Existence and Attributes of the Deity*.

In the opening chapters of *Natural Theology*, Paley invites the reader to imagine walking across a meadow and finding a stone. In an insult to geologists that still stings two hundred years later, Paley says that the stone is not particularly interesting because there is little to be learned about that stone or its history. However, Paley explains, if you were to find a watch in the meadow, then you have found something interesting. The intricacy and complexity of the watch immediately let you know that somewhere out there is a watchmaker. Such precision and purpose could not possibly come about naturally. Furthermore, by careful examination of how well the watch is made, you can infer how good at their craft the watchmaker is.[13]

According to Paley, you can then extend this argument to the meadow itself. After all, a meadow is full of flowers, bees, squirrels, and any number of other living things, each of which is just as complex and intricate as any watch. Moreover, all of these individual occupants of the meadow fit together perfectly to create a balanced and functioning system we call nature. If a watch implies a watchmaker, then nature implies a naturemaker. All of the perfectly harmonious interconnections we see in nature let us know that the maker is more than just competent: the maker is good.[14]

The argument Paley makes in *Natural Theology* is an old one. It goes back to Aristotle, but its most famous proponent, prior to Paley, was the theologian Thomas Aquinas in the thirteenth century. In his *Summa Theologica* (written in the 1200s but not published until the 1400s), Aquinas set out to demonstrate that you could prove the existence of God purely through reason. He established five arguments for the existence of God, one of which is the argument from design—that the appearance of direction and purpose in nature comes about because there was purpose in its design by a designer.[15]

Paley's book was well received by the British public, and its contents became folk wisdom. One contemporary review stated, "No thinking man, we conceive, can doubt that there are marks of design in the universe."[16] Conservative quarterly *The British Critic* put its review of Paley on its front page, calling it a "new treasure," and declared that, while the arguments weren't necessarily new, the structure of the book and the depth of its insight made its conclusions irrefutable.[17] This, then, is the world that Charles Darwin was born into—a world in which the good design of nature was inextricably linked to the goodness of God, and in which the transformation of species was an entertaining, but easily dismissed, idea.

The Origin of *The Origin*

Charles Robert Darwin (Bobby to his family) was born in 1809. His father was a successful doctor and investor named Robert Waring Darwin.[18] His mother was Susannah Wedgewood of the Wedgewood pottery family.[19] Given his background, Charles could have lived a comfortable life of leisure. In fact, his older brother, Erasmus Alvey "Ras" Darwin, retired from medicine after a one-year career at the age of twenty-nine. Ostensibly, this was because his father was concerned that he was too frail for the life of a doctor. Since Ras Darwin lived another fifty-two years after retirement and was rumored to be romancing several different women (including his own cousin's wife),[20] the retirement may have been more of a lifestyle choice than strictly health-related. It's also possible that at 6'3" and 330 lbs., Robert Darwin was a poor judge of what constituted frailty.[21]

Like his older brother, his father, and his grandfather before him, Charles Darwin's original plan was to become a doctor. But unlike his ancestors, Charles discovered during his medical school years that he didn't like it. At his father's insistence, Charles moved from Edinburgh to Cambridge and began an educational path toward the clergy. In his autobiography, Charles chalks that decision up to his father, who was "very properly vehement against my turning an idle sportsman, which then seemed my probable destination."[22] To be clear, Darwin's sudden interest in the clergy wasn't necessarily the result of some inspired calling. Although he did describe himself as "quite orthodox" on matters of religion at that point in his life,[23] the clergy was simply a standard career path in the 1800s for second sons of prominent families. It was a way to make an honest living without necessarily needing to work too hard at it.[24]

Darwin's ecclesiastical career never materialized. He was, however, always interested in nature, and while at Cambridge, he spent many of his free nights among the geologists and biologists there. In 1831, one of his mentors, John Stevens Henslow (himself both a botanist and an Anglican priest), presented him with a phenomenal opportunity. Henslow had been asked to find a naturalist to join the crew of the *HMS Beagle*, a navy ship that would be conducting a two-year survey of the coast of South America. One unusual aspect of the search was that the captain was looking for a man who could be recommended both as a naturalist and as a gentleman. Henslow told Darwin that he was the perfect man for the job.[25]

Charles Darwin was immediately excited by this opportunity. Robert Darwin was less so. In fact, he initially forbade his son to accept the position. He did, however, say that if Charles could find any sensible person who could give a good reason for the voyage, he would reconsider. That sensible person turned out to be Charles's uncle, Josiah Wedgewood, and that good reason was that it would be exceedingly difficult for Charles (who had a reputation for throwing money around at school) to spend much money at sea.[26] Within a week, Charles had his father's blessing and had secured the position.

What started out as a two-year voyage on the *Beagle*, eventually grew to five years. Through it all. Darwin kept a diary of his life and experiences.[27] No detail was too mundane to be recorded. In five years of his *Beagle* diary, Darwin documents the minutia of his days down to the shopping trips, seasickness, and sermons. But never once does he mention evolution, transformation, or anything else that would be relevant to his eventual evolutionary insight.[28] Darwin's trip on the *Beagle* allowed him to hone important skills for a naturalist, including specimen collection, observation, and classification. It provided him with experiences that would prompt him to think about natural history in new ways. But he didn't put everything together until years after his return to England.

Darwin's evolutionary insight came only after he paired his experiences on the *Beagle* with readings he had done in other fields, especially economics. The thoroughness of Darwin's diary-keeping lets us know exactly when he had his ideas and also what he was reading at the time. In 1838, Darwin sketched his first evolutionary tree in his scientific journal underneath the phrase "I think. . . ."[29] At the same time, his reading journal lets us know that he was reading Thomas Malthus and Adam Smith, two economists whose works specifically emphasized the role of competition for limited resources in the shaping of history and society.[30]

Darwin's Insight

In *On the Origin of Species*, Charles Darwin did two things. He established the fact of evolution; that is, he documented that evolution is an observable phenomenon. He also proposed a theory of evolution; that is, a mechanism by which evolution might occur.

To demonstrate the factuality of evolution, Darwin relied on three main lines of evidence, beginning with the easily observable changes that can occur in species under domestication. Today, just a look around the grocery store will confirm that you can selectively breed for new traits. Whether you're buying a seedless watermelon or a 2 lb. chicken breast, you are buying something that never existed in nature. Darwin concentrated his attention on pigeons. Pigeons were a convenient and accessible example of domesticated varieties for Darwin, as raising pigeons was a popular hobby in Victorian England. The first chapter of *On the Origin of Species* documents all of the different varieties of pigeon that breeders had created from selective breeding of the common rock pigeon.

In his second line of evidence, Darwin extends the idea of change in species over time by examining the fossil record. Darwin was writing shortly after William Smith had put fossils into their proper time order, and everyone could see that the further back in time you go, the more different fossils became. Darwin attributed this to nature acting on life the same way that a pigeon breeder does—maintaining some traits and eliminating others. But unlike a breeder, who only works over a few generations, nature has acted for millions of years and can produce bigger changes.

Darwin's most compelling line of evidence for evolution was also his most revolutionary because it flew in the face of Paley's argument in *Natural Theology*. Darwin points out that some things are designed so badly that they can only be explained through historical accident. One particularly visceral example of questionable design choice, although it is one that Darwin did not use, is the human throat (or, as biologists call it, the pharynx). It's the pipe that air uses to get to the lungs. If it gets clogged, death can occur in as little as five minutes. However, the pharynx is also the pipe that food uses to get to the stomach. Using the throat as a passage for both air and food is a terrible idea. A kind creator building humans from scratch would probably come up with some alternative arrangement that made choking less likely. The design makes perfect sense, however, if you understand that all modern vertebrates have evolved from fish, who use their pharynx to eat, but breathe through their gills. Once vertebrates began to move onto land, the pharynx was an existing pathway from the outside world to the inside world that was readily available for use as an airway with little modification. It was a convenient happenstance that the pharynx was adaptable to breathing air, but not an optimal one.

After documenting that evolution has happened, Darwin moves on to proposing *how* it happened. There are a lot of complex, nearly impenetrable theories in science. Charles Darwin's theory of evolution by natural selection is not one of them. In 1887, the biologist Thomas Henry Huxley famously wrote, "My reflection, when I first made myself master of the central idea of the 'Origin,' was, 'How extremely stupid not to have thought of that!'"[31] It is three simple facts tied together to draw a conclusion.

1. All populations vary. In some species, this is more obvious than others. All squirrels may look alike to us, but they are not. Squirrels vary in their size, coloration, fuzziness, and in literally thousands of other ways at the genetic level.
2. Traits are inheritable. At the time, nobody knew why this was true, but it was a basic fact of life. Gregor Mendel was experimenting with pea plants and discovering the foundations of genetics at about the same time that Darwin was writing *On the Origin of Species*, but nobody knew about it yet. They did know, however, that if your kids look more like your neighbors than they look like you, somebody has some explaining to do.
3. Some creatures die young. The word for this in Darwin's day was *superfecundity*. All it means is that more creatures are born than will live to reproduce. It's a sad fact of life, but it is a fact of life.

Natural selection says that the variation in a population will be one factor that determines who lives to reproduce and who does not, and that the traits that help you to survive to reproductive age will be passed to your offspring. That's it. It's simple and straightforward, almost to the point of folk wisdom. And for a while, it was a crime to teach it in seven states.

What's All the Fuss About?

Many myths about Darwin unravel on close inspection. One example is the myth that Darwin based his theories primarily on his observation of Galapagos finches. That story is a scientific myth akin to Newton's apple. The word *finch* appears exactly twice in *On the Origin of Species*: Darwin describes a pigeon as having a beak about the same size as a finch's beak. Later, he notes that it is possible to cross a finch with a canary.[32] His one mention of the Galapagos is a bit more relevant to evolution. As part of a discussion of how botanists disagree about the number of genera of plants in Britain, Darwin remarks that he had a similar problem with birds in the Galapagos.[33] In his introduction, Darwin explicitly states that while his theory helps to explain things that he saw on his voyage around the world, the theory itself came much later.[34]

Another myth is that his theory of evolution caused an immediate religious backlash in America. However, Darwin published *On the Origin of Species* in 1859, and the first anti-evolution law in the United States (Tennessee's *Butler Act*) wasn't drafted until 1925. To put that time gap into perspective, Warren G. Harding was born in 1865 into a world where nobody really questioned the teaching of evolution. He had a successful career in politics culminating in being elected the twenty-ninth president of the United States, and then he died in 1923 in a world where nobody still really questioned the teaching of evolution. The backlash against Darwin and evolution did eventually come, but it was hardly immediate and, at least at first, was not religiously motivated. Ironically, the anti-Darwin movement started out as a response to some things that Darwin never wrote.

There are some surprising words and phrases missing from the first edition of *On the Origin of Species*. First among them is *evolution*. Evolution has become synonymous with the process that Darwin describes, but he himself didn't use that term in his original work. The word existed in the 1850s and had several different meanings related to turning or unfolding. A naval fleet could perform an evolution to reposition itself. The evolution of time could reveal new facts. A person could watch the evolution of a butterfly from its cocoon.[35] However, calling the descent and modification of life from a single common ancestor to its modern diversity *evolution* wouldn't have made any sense to Darwin. Instead, he called it *descent with modification.*[36]

Another term absent from the first edition of *On the Origin of Species* is *survival of the fittest.* This term was coined five years later by the philosopher Herbert Spencer.[37] Darwin didn't coin the term, but he did like it and began to use it in later editions of *On the Origin of Species.* In a letter he wrote to the biologist Alfred Russel Wallace in 1866, Darwin remarked ironically that there were now two terms for his mechanism, *natural selection* and *survival of the fittest,* and which one would catch on would come down to a matter of the survival of the fittest.[38]

While this remark to Wallace was meant to be a small joke, it was actually quite prescient. Notice that Darwin didn't say that it would come down to a matter of natural selection. That wouldn't have made as much sense. Nature wouldn't be doing the selecting. People would. As opposed to the term *natural selection, survival of the fittest* implies broader applicability. After all, natural selection tells you right away that it is a theory about nature. In contrast, survival of the fittest can apply anywhere. Either term might explain how a peacock acquired its remarkable tail, but survival of the fittest can justify marketplace monopolies and playground bullying as well. As it turned out, that's exactly what happened. The term *survival of the fittest* became coopted to explain human affairs and became a justification for cruelty.

Social Darwinism is the doctrine that we should let our marketplaces work the same way that nature does—purely through a struggle for existence and survival of the fittest. If a company fails, the working class struggles, or a person loses their job, that will improve the state of the economy in the long term. The doctrine says that trying to prevent any of those things from happening isn't just counterproductive, it's unnatural. It is *laissez-faire* capitalism dressed up in the trappings of science. "Social Darwinism" is misnamed. Darwin himself never advocated for anything remotely like it. It grew out of Spencer's writings.

The relationship between Darwinian evolution and free-market capitalism is a complicated one. The heart of Darwin's argument is so similar to the argument that Adam Smith makes in *The Wealth of Nations* that evolutionary theorist Stephen Jay Gould once described them as "Isomorphic."[39] That is, the two theories are identical except for their subject matter. Each theory argues that competition between individuals for limited resources will result in a balanced system. Unsuccessful individuals will tend to fail over time (for Smith, this is business bankruptcies; for Darwin, species extinctions), and successful individuals will continue to prosper in the future.

There is, of course, a significant difference between nature and commerce. Commerce is a human invention and is governed by human rules and morality. Success is generally cheered. Failure in business is sad, but there is opportunity for redemption in your next venture. Nature, at least to Darwin, has no such moral underpinnings. The extinction of a species isn't good or bad, tragedy or triumph. It is just part of the process; and, in fact, it is a fate that has befallen the overwhelming majority of species that have ever existed. No species (at least in pre-human history) ever went to any great lengths to save another from that fate.

In addition to being misnamed and morally reprehensible, Social Darwinism is also based on a logical fallacy, the belief that things we deem to be natural are inherently good.[40] This naturalistic fallacy is closely related to another fallacy that Scottish philosopher David Hume identified as the is-ought fallacy, in which the way things are is the way things ought to be.[41] Like the power of artificial selection, the naturalistic fallacy is perhaps best illustrated by a trip through the grocery store. Every label beckoning you with claims of "all-natural," "no artificial ingredients," or "made with real sugar" is an invitation to confuse natural with good. Presently, it is the mindset behind anti-vaccination practices, formula-shaming of new mothers, and dozens of other societal ills. In the guise of Social Darwinism, the naturalistic fallacy has been used historically to justify poverty, abuse, colonialism, and genocide.

By the 1920s, Social Darwinism had become widespread enough a doctrine that a political movement arose to try to limit its influence. The easiest way to stop the spread of Social Darwinism, they reasoned, was to stop people from learning about Darwinism in the first place. This movement for social reform would eventually become the foundation for American creationism, and it all began with one of the great social reformers of the time.

The Great Commoner

When William Jennings Bryan was only thirty-six years old, he became the youngest person ever to win electoral votes for president, a record he still holds. Undeterred by his loss to William McKinley, Bryan ran for president twice more in 1900 and 1908. While he never became president, he did earn a total of 494 electoral votes across his three attempts—more than any other person in American history who never actually became president. His ability to connect with everyday rural Americans earned him the nickname "the Great Commoner." To make a long story short, William Jennings Bryan was one of the most popular and beloved politicians of the early twentieth century. Bryan gave up on his presidential aspirations after 1908 but stayed active in politics, eventually becoming Woodrow Wilson's secretary of state. As someone immersed in international politics in the early twentieth century, Bryan had a front-row seat for the rise of Social Darwinism in Europe and grew concerned about the societal consequences at home if too many Americans began to learn about evolutionary theory.[42]

The first time Bryan spoke publicly about evolution was in a 1909 revival speech titled "The Prince of Peace." In the speech (delivered fifty years after *On the Origin of Species* was published), Bryan states that he has serious doubts about Darwin's theory:

> I do not carry the doctrine of evolution as far as some do; I am not yet convinced that man is a lineal descendant of the lower animals. I do not mean to find fault with you if you want to accept the theory; all I mean to say is that while you may trace your ancestry back to the monkey if you find pleasure or pride in doing so, you shall not connect me with your family tree without more evidence than has yet been produced. I object to the theory for several reasons.[43]

Bryan then enumerates his objections, and they fall into two broad camps. First, he is concerned that if humans think of themselves as related to monkeys, they may lose sight of their link to the divine. Second, he describes Darwin's theory, with its reliance on struggle and competition, as a "law of hate," and divine creation as the "law of love."[44]

Bryan made it clear that he saw the amorality of natural selection as its greatest flaw. In a speech that he wrote shortly before his death but never got to deliver, he wrote, "Science is a magnificent material force, but it is not a teacher of morals. It can perfect machinery, but it adds no moral restraints to protect society from the misuse of the machine."[45]

Interestingly, the only time Bryan ever really addressed any of the actual evidence for evolution was a passage in "The Prince of Peace" in which he freely acknowledged that there is some physical similarity between humans and other primates. His objections were not to the content of the theory or the evidence for it. They were to the ramifications of the theory. This is a type of logical fallacy called the appeal to consequences fallacy. In an appeal to consequence, you judge a premise not based on the evidence for or against that premise but based on the morality or immorality of its consequences.[46] The structure goes like this:

- If A is true, then B must be true.
- It would be awful if B were true.
- A must not be true.

You can also commit this fallacy in a positive form:

- If A is true, then B must be true.
- Gee, it would be nice if B were true.
- A must be true.

When you phrase the appeal to consequences fallacy this way, it becomes clear that it is really just a form of wishful thinking. In Bryan's case, the argument was that if natural selection is correct, then it can be used to justify Social Darwinism. Social Darwinism is awful, so evolution must be wrong.

This line of reasoning is illustrated by another passage from his final intended speech. Bryan was citing the recent publication of an academic journal article that suggested that periods of societal advance are often associated with a loosening of sexual taboos. Outraged by this advocacy of animalistic behavior, Bryan wrote:

> No one charges or suspects that all or any large percentage of the advocates of evolution sympathize with this loathsome application of evolution to social life, but it is worth while to inquire why those in charge of a great institution of learning allow such filth to be poured out for the stirring of the passions of its students.[47]

Bryan didn't care if evolution was right or wrong. He also didn't care what percentage of evolutionists were actually Social Darwinists. His only concern was for the consequences of evolutionary thinking. He was concerned that it would undermine people's faith in the Bible and be used to justify cruelty and immorality in public policy.

By the 1920s, Bryan's political career was largely over, but his political influence still loomed large. His decades of denouncing evolution began to have legal consequence. Several southern states began restricting the teaching of evolution. In some states, including Georgia and Texas, this was achieved through school board proclamations. In Florida, the state legislature issued a proclamation expressing disapproval of evolution.[48] In 1925, Tennessee was the first state to outright ban evolution from public classrooms through passage of the Butler Act. The Butler Act was simple and straightforward. No teacher in any state-funded school could teach any theory that deviated from the Biblical account of creation under penalty of law.

There is little question that the Butler Act was inspired by Bryan's preaching. In 1925, Tennessee attorney W. B. Marr, one of the bill's architects, wrote to Bryan saying,

> We feel that we are almost the proximate cause of this statute in that we heard you present your great lecture "Is the Bible True?" Later we had several thousand of them published and distributed generally. Later when the Legislature first convened we sent about 500 copies to the members. Evidently this caused Mr. Butler to read and think deeply on this subject and prompted him to introduce his bill.[49]

Bryan himself even had a small hand in the bill's content. He advised the Tennessee legislature to take a clause out of an earlier draft specifying a penalty for noncompliance, as that might give opponents of the bill a means for objection.[50] That turned out to be a moot point. The Butler Act passed its final vote in the Tennessee House with a vote of 71 yeas, 0 nays, and 5 presents.[51] It would inspire several other states to pass anti-evolution laws of their own.

Anti-Evolution Laws

The story of how the Butler Act was challenged is a famous one—the so-called Scopes Monkey Trial held in Dayton, Tennessee, in 1925. The defendant, John Scopes, became the first person in American history put on trial for teaching Darwin's theory of natural selection. It was also far more than just a trial. It became one of the first true media fiascoes. People swarmed in from all over the state and even all over the country to watch the proceedings. It was the first trial broadcast by radio. It became the basis for the play *Inherit the Wind*.[52] Bryan himself assisted at the prosecutor's table, and the American Civil Liberties Union (ACLU) sent their best attorney, Clarence Darrow, to defend Scopes.

In many ways, the actual Scopes case wasn't really worth the hoopla of the trial. John Scopes wasn't some radical firebrand defiantly challenging an unjust law. He was the local football coach who did occasional work as a substitute teacher.[53] Scopes didn't teach evolution to cause trouble. In 1925, Tennessee had two contradictory laws on the books. The Butler Act forbade the teaching of evolution. At the same time, state school standards required the use of a textbook that not only promoted evolution but also was written by a Social Darwinist.[54] Teaching evolution was illegal, but so was not teaching evolution.

The case of *Scopes v. State* may have built to a crescendo, but it started quietly enough. Despite how the trial has been portrayed in later media, Scopes was not particularly demonized for teaching evolution. Many people were happy that he had done it, as it brought attention to the town. In the 1920s, Dayton, Tennessee, was a struggling coal town looking for a way to attract attention and tax revenues. When the town fathers learned that the ACLU was looking for a volunteer to serve as a test case to challenge the Butler Act, they asked Scopes if he was willing to be arrested. After some initial reluctance, he acquiesced.[55]

The trial of John Scopes lasted for nearly two weeks. It included a day of outdoor proceedings due to excessive heat, and the defense team calling William Jennings Bryan (who was co-counsel for the prosecution) as a witness on the historicity of the Bible.[56] After all the fuss, Scopes was found guilty—because he was guilty. Nobody ever questioned that little fact. Scopes had indeed taught evolution in defiance of the Butler Act, and the judge fined him $100.[57]

This should have been the beginning of the end for anti-evolution laws in America. Scopes had been convicted. Now the ACLU could appeal and re-appeal his conviction until they got to the Supreme Court, who, they assumed, would overturn the Butler Act for violating the First Amendment. Unfortunately, the case never made it that far. At the very first appeal, the Tennessee Supreme Court threw out Scopes's convictions on procedural grounds. In Tennessee in 1925, only a jury could issue a fine of over $50.[58] Scopes's conviction was unappealable, and it would be almost another half-century before anti-evolution laws would come before the Supreme Court.

Other states followed Tennessee's lead in outlawing the teaching of evolution, including Arkansas, which, in 1928, passed its own law barring Darwin from the classroom.[59] But in 1965, just like in Tennessee a few decades earlier, the Little Rock school board adopted a new required biology textbook that included a chapter on evolution.[60] Once again, high school biology teachers were placed in an unwinnable situation. Teaching evolution was illegal, but so was not teaching evolution.

The Arkansas anti-evolution law did eventually make its way to the Supreme Court. Little Rock biology teacher Susan Epperson filed suit against the state seeking an injunction guaranteeing that she wouldn't be fired for following the required curriculum. In 1968, Justice Abe Fortas wrote on behalf of a unanimous Supreme Court that anti-evolution laws were based on an inherently religious agenda and that a state cannot pass a law "hostile to any religion or to the advocacy of no-religion, and it may not aid, foster, or promote one religion or religious theory against another or even against the militant opposite."[61] Forty-three years after the passage of the Butler Act, anti-evolution laws were finally struck down.

Although the anti-evolution movement of the 1920s may have had its origins in social reform and a desire to avoid the slippery slope to Social Darwinism, it was also rooted in religious fundamentalism. By crafting laws that specifically prohibited departure from scriptural authority, anti-evolution legislators were aligning themselves with the doctrine of creationism—the idea that all of nature was recently created in its present state as outlined in Genesis. These laws were doomed to inevitable failure when they reached the Supreme Court, as Justice Fortas's ruling makes clear. However, the anti-evolution movement didn't go away in the wake of this judicial defeat. Ironically, it adapted.

Intelligent Design

The modern anti-evolution movement in America primarily centers around a theory called Intelligent Design (or ID). A lot of people misunderstand ID, thinking that it is a new spin on the creationist movement of the 1920s. It's not. There are many similarities between the two theories, but in some ways, creationism is a more respectable doctrine than ID. Neither theory can correctly be called scientific, but creationism is more open about its particular research agenda than ID and adheres better to the rules of honest discourse than ID theory does.

Like creationism, ID also promotes the idea that the universe was created in its present form by (as the name implies) some intelligent designer. And like creationism, ID proponents want to see their ideas taught in science classes around the country. However, because of the legal history of creationist and anti-evolution legislation, ID has to deny what it truly is. If it openly promoted a theistic agenda, then it would be in violation of established legal precedent. Instead, ID proponents try to poke holes in evolutionary theory and then sell their theory as a scientific patch for these holes.

Intelligent Design argues that there are complex aspects of our universe, and particularly of biology, that show evidence of intention and purpose. It further argues that the only possible origin for this apparent complexity, intention, and purpose is through design. Something as complex as an eye, or DNA, or a bacterial flagellum, could not have originated through a process of natural selection acting on individual mutations and variability. Such complex structures could only have come about through the action of a designer. Moreover, some naturally occurring systems (like DNA) convey information, and information can only come about through the action of an intelligence. If that sounds familiar, that's because it is exactly the argument that William Paley made in *Natural Theology* more than two hundred years ago. The intricacy of a watch, combined with the watch's utility as a timepiece, is incontrovertible evidence of a watchmaker.

The Supreme Court's ruling in the Epperson case makes it clear that religiously motivated curricula may not be taught in public schools, so ID theorists are very careful to distance themselves from any particular religion. According to the website for one of America's largest centers for ID research, the Discovery Institute, ID is not creationism because it is "agnostic" about exactly who the designer is. It's not necessarily the God of Genesis.[62] It is simply an unknown being (or beings) who designed and created the universe. This argument is particularly hard to swallow (down our poorly designed pharynx) when you remember that before Paley, the initial popularizer of the argument from design was St. Thomas Aquinas, who specifically crafted the argument to prove the existence and nature of the Biblical God. ID theory claims that you can see the design in nature through the purposeful action of ecosystems, organisms, organs, and even individual molecules. Or, as Aquinas put it,

> We see that things which lack intelligence, such as natural bodies, act for an end, and this is evident from their acting always, or nearly always, in the same way, so as to obtain the best result. Hence it is plain that not fortuitously, but designedly, do they achieve their end. Now whatever lacks intelligence cannot move towards an end, unless it be directed by some being endowed with knowledge and intelligence; as the arrow is shot to its mark by the archer. Therefore some intelligent being exists by whom all natural things are directed to their end; and this being we call God.[63]

ID proponents like to sell their theory as a new and cutting-edge rebuttal to Darwin. ID critics often refer to it as "creationism in a cheap tuxedo,"[64] implying that it is the same old theory trying to pass itself off in more respectable trappings. In reality, it is neither. It is the medieval theology of St. Thomas Aquinas.

Creationism works from the assumption that the Bible is correct and that any alternative must be based on faulty premises. You can accurately call it a philosophy, a doctrine, or even a theology, but it does not qualify as science. Creationism isn't science because it doesn't follow the most fundamental assumption of science that originated with Thales (in chapter 1): nature is a result of constant knowable rules, not the unpredictable actions of capricious agents. This is also where ID fails as science. It doesn't matter whether or not their designer is the God of Genesis or not. Whether they call their designer God, Brahma, Gaia, or just the Designer, by relying on the actions of an intelligent being as a mode of explanation, ID theorists place their doctrine outside the realm of science.

Another similarity between creationism and Intelligent Design is their reliance on the argument from ignorance to promote their agenda. In fact, the argument from ignorance is central to ID in that ID's entire premise is that there is no way to explain certain phenomena without relying on a designer. One classic ID argument is that the bacterial flagellum (essentially the tiny propeller on the back of a single-celled organism) could not possibly have originated through natural selection. The Discovery Institute's FAQ page links readers to an essay by one of the institute's founders asserting that since the flagellum requires the interaction of more than thirty different proteins, "natural selection can 'select' the motor once it has arisen as a functioning whole, but it cannot produce the motor in a step-by-step Darwinian fashion."[65] In 2007, the National Academy of Sciences published an article in their proceedings titled "Stepwise Formation of the Bacterial Flagellar System," explaining the precise step-by-step genetic history that led to the evolution of the flagellum by natural selection.[66] But fifteen years later, the Discovery Institute is still claiming that nobody can explain it.

Legally, you can't bar teachers from teaching evolution. You also can't require teachers to teach an overtly religious theory. ID advocates claim that their argument is not a religious one, but courts have ruled otherwise. Most notably, in the 2005 case *Kitzmiller v. Dover*, Judge John E. Jones III ruled that ID could not be a required part of the science curriculum in Pennsylvania schools, since its reliance on intelligent agency is not scientific. So more recently, ID advocates have turned to a new strategy. If they can't promote their own agenda, they can actively work to undermine the scientific alternative. ID theorists continue to argue from ignorance so that they can claim a controversy exists where there is none. They then work to "empower" teachers to "teach the controversy."

"Academic Freedom Acts" are state laws that provide school teachers the freedom to present evidence against any scientific principle that they deem controversial. The state may require them to teach Darwin, but it can't bar them from then spending time on how evolution can't explain (for example) the bacterial flagellum. Many of these state laws are modeled after a 2001 amendment to the conference report for the No Child Left Behind Act called the "Santorum Amendment." The amendment states, in part, that "where topics are taught that may generate controversy (such as biological evolution), the curriculum should help students to understand the full range of scientific views that exist, why such topics may generate controversy, and how scientific discoveries can profoundly affect society."[67] This amendment was proposed by then-senator Rick Santorum of Pennsylvania, but it was co-authored by Phillip Johnson of the Discovery Institute.[68]

Evolution and Faith

A false dilemma fallacy occurs when you mistakenly believe that you need to choose between two alternatives when those options are not, in fact, mutually exclusive.[69] The belief that there is an inherent conflict between evolution and faith is a classic example of this fallacy. It is perfectly possible to believe in evolution and God simultaneously. If you don't believe me, read your Darwin. The word *evolution* does not appear in *On the Origin of Species*. The phrase *survival of the fittest* only appears in later editions. But the word *creator* appears in every edition. For example,

> Authors of the highest eminence seem to be fully satisfied with the view that each species has been independently created. To my mind it accords better with what we know of the laws impressed on matter by the Creator, that the production and extinction of the past and present inhabitants of the world should have been due to secondary causes, like those determining the birth and death of the individual.[70]

As a matter of fact, the word appears more in later editions than earlier editions. Darwin added a reference to the creator into the sixth edition (published in 1872, when Darwin was in his sixties). The very last sentence of that edition begins, "There is grandeur in this view of life, with its several powers, having been originally breathed by the Creator into a few forms or into one." Previous editions had simply said "breathed into a few forms."[71] Darwin saw no conflict between his theory and his faith.

Darwin's own religious beliefs evolved (for lack of a better word) over his lifetime. He reflects on this evolution in his autobiography, saying that while he was on board the *Beagle*, other sailors would make fun of how often he quoted scripture.[72] By the end of his life, however, Darwin had mostly abandoned his faith. But there was no sudden moment of revelation where Darwin realized that evolution disproved the existence of God or His active role in the universe. In fact, his break with Christianity came not from a conflict between science and faith, but a conflict between his desire to believe and the church doctrine of the time. He describes it as a slow creeping doubt and says that the final straw came not from anything in nature or the Bible, but from one particular aspect of dogma:

> I can indeed hardly see how anyone ought to wish Christianity to be true; for if so the plain language of the text seems to show that the men who do not believe, and this would include my Father, Brother and almost all my best friends, will be everlastingly punished. And this is a damnable doctrine.[73]

Charles Darwin rejected Christianity for the same reason that William Jennings Bryan rejected evolution. It wasn't out of any inherent conflict between science and faith, but because he didn't like the consequences of believing.

CHAPTER 9

The Map Changes

The theory of plate tectonics is the framework within which all geologists work. It explains the distribution of continents, oceans, volcanoes, earthquakes, fossils, and rocks. Unlike many scientific theories, there is no particular person who is credited with discovering plate tectonics. It was put together based on the work of hundreds of geologists in dozens of countries. Also, unlike evolution, climate change, heliocentricity, or any of the other topics in this book, there is no organized anti-tectonics movement in America. But there used to be. The idea that large chunks of the earth's crust could move across the surface of the globe was once strongly opposed and viciously attacked—by geologists.

The Mythical Land of Dr. Suess

Trees can't swim. That may seem obvious, but it's important to keep that obvious fact in mind when considering the work of Israel C. White. After the Civil War, White was West Virginia's first official state geologist and became one of the preeminent experts on coal in his time. White was such a noted coal geologist that in 1904, when the Brazilian government wanted to evaluate the economic potential for coal development in their country, they had White lead the team. It was while working in Brazil that White discovered his problematic trees.

The trees in question belong to a genus called *Glossopteris*. They are extinct today, but in the Permian (a period of geologic time that stretches from about 300 million to 250 million years ago), they were abundant. Their fossils are very typically found in coal-bearing horizons like the ones White was studying in Brazil. But White knew that they were also common in coal-bearing deposits of the same age in South Africa.[1] *Glossopteris* had apparently crossed thousands of miles of open ocean. And that's a problem because trees can't swim.

Even before White's discovery, *Glossopteris* fossils were already well known for having a geographic distribution that was difficult to explain. They are found in several parts of the southern hemisphere that are not connected by land. In 1885, the Austrian geologist Eduard Suess proposed an explanation for this strange distribution: The Indian Ocean, he claimed, was missing a continent. Suess proposed that eastern Africa, Madagascar, and India were the remnants of a larger continent, part of which had sunk into the sea. He called this missing continent Gondwana.[2]

Suess was not the first person to propose a missing continent in the Indian Ocean based on an unusual distribution of fossils. Today, wild lemurs are only found on the island of Madagascar. But fossil lemurs can be found in India. Strangely, there are no fossil lemurs in Africa, the Middle East, or any of the other areas that a lemur would need to walk through to get from India to Madagascar. In 1864, the British zoologist Philip Sclater tried to explain this strange distribution by inventing the sunken continent of Lemuria.[3]

Gondwana and Lemuria may sound fanciful today, but, by the geologic standards for the time, they were reasonable hypotheses. The sedimentary rock record showed ample evidence that sea levels had changed through time and the regions that geologists were trying to connect weren't that far apart geographically. Gondwana or Lemuria would only have extended the coastlines of the western Indian Ocean by about a thousand miles. But finding *Glossopteris* in South America was a problem. It's nearly five thousand miles from South Africa to Brazil, with the Atlantic Ocean in between. So how did the same group of trees end up populating both regions hundreds of millions of years before people walked the earth? In White's day, the best explanation geologists had was land bridges.

You may remember the idea of land bridges from an elementary school social studies class about how the first people made it to the Americas. The idea is simple enough. At some point during the last ice age, the combination of sea level drop and sea ice formation created a walkable path across the Bering Strait, making it possible to walk from Siberia to Alaska. But the Bering Strait is only a few hundred feet deep, only about fifty miles wide, and sits at a latitude prone to freezing. Making that path walkable seems far more plausible than an unknown land bridge across five thousand miles of tropical ocean with an average depth of more than two miles.

Geologists called the hypothesized land bridge connecting Brazil to South Africa Archhelenis. This was, of course, to distinguish it from the numerous other land bridges that they believed had crisscrossed the ancient oceans. Archatlantis connected the Caribbean to Northern Africa. Archinotis connected South America to Antarctica. Most impressive of all was Archigalenis, which connected Peru to Siberia—a distance of nearly ten thousand miles.[4] Geologists at the time said all of these land bridges had sunk into the ocean long ago, but you could still see bits and pieces of them sticking up above the surface. Some geologists believed that the Hawaiian Islands (which form a southeast-to-northwest chain) were the mountains of the now sunken Archigalenis.[5]

Understanding how such large features of the earth could sink into the ocean requires understanding how geologists of the time were thinking about the earth and its history. The prevailing wisdom was that the earth's history was primarily a history of cooling. The argument was that the earth started as a ball of molten rock in the cold vacuum of space and has been cooling ever since. The crust is solid, because it is the outer layer and cooled first; however, the inner earth is still cooling. As things cool, they contract. So the interior of the earth is still contracting and the crust occasionally crumples due to this inner contraction.[6] This crumpling causes some areas to fold outward, creating mountains, and other areas to fold inward, creating ocean basins.

The idea that a cooling, contracting, and crumpling earth could account for the major topographic features of the continents held sway for most of the nineteenth and early twentieth centuries. It also made it clear that the only way any chunk of the crust moves relative to its neighbors is vertically. Continents could rise, and continents could sink. But absolutely nothing moved horizontally across the surface of the earth. It would take well into the twentieth century before anyone suggested that geologists were thinking in the wrong direction and until remarkably recently before anyone really took the idea seriously.

Continental Drift

The twentieth century saw a long, slow shift in how geologists envisioned the motion of landmasses. At the beginning of the century, the prevailing wisdom was that landmasses could only move vertically. The person who would present the first major theory that continents moved horizontally initially became famous for an extraordinary feat of verticality. In 1906, Alfred Wegener, along with his older brother Kurt, set a world record for the longest time aloft in a hot air balloon, fifty-two hours.[7] This feat wasn't purely thrill-seeking. The Wegener brothers studied weather and climate by using balloons and kites to make atmospheric observations. Alfred in particular was fascinated by arctic climates and undertook several expeditions to Greenland to measure its ice caps.

Alfred Wegener is an interesting character in that nobody seems to agree on exactly who he was. He was fascinated by climate, but there was no such word as *climatologist* in his day. Most sources call him a meteorologist. Some call him an explorer. Some just call him an adventurer. All of these are true to some extent, but this broad résumé is important for two reasons. First, it means that Wegener was well-traveled, particularly among the far northern regions of Europe and North America. Second, the one thing everyone seems to agree on is that Wegener was not a geologist[8]—a fact that would become very important.

In 1912, Wegener first published his theory of continental drift.[9] Instead of invoking land bridges or sunken continents to explain similarities across the oceans, continental drift suggested that the continents had once all been connected into one large landmass that he called Pangea. He further claimed that Pangea had broken apart and that the fractured remains (our modern continents) spent the recent geologic past moving away from one another. Continental drift was based on a number of observations that Wegener had made, beginning with the remarkable symmetry of coastlines on either side of the Atlantic.

Wegener was not the first person to notice that the coastlines of the Americas mirror the coastlines of Europe and Africa. People began to notice that almost as soon as the first good maps of the new world had been made. As early as 1596, the Dutch cartographer Abraham Ortelius suggested that the Americas had once been a part of the Old World but had been torn away from the other continents by violent storms.[10] Wegener did more than just match up the edges of continents. He was able to establish that rock types, mountain chains, and fossil ranges matched up across great distances as well. The fossil range of *Glossopteris*, for example, becomes much more sensible if you envision it living in southern Pangea instead of across multiple distinct continents. (See figure 9.1.)

Wegener also specifically demonstrated what he perceived as a fatal flaw in the theory of land bridges. The difference in density between the continental and oceanic parts of the crust made the sinking of continents nearly impossible. In one fell swoop, Wegener provided an explanation for the distribution of rocks, fossils, and shorelines, and demonstrated the physical impossibility of the prevailing model. Yet Wegener's model was never really embraced by the geological community. It was rejected as a plausible model for several reasons—some good, some bad.

Continental drift turned out to be more wrong than right, but not really in any way that geologists could demonstrate at the time. Continents have, in fact, moved across the surface of the earth—but not at all in the way that Wegener had envisioned. Wegener believed that the continents plowed their way through an ocean floor that yielded to their intrusion like a viscous fluid.[11] That doesn't happen. Oceanic crust is every bit as rigid as the continental crust. Of course, in 1912, nobody could really be certain of that fact. Our knowledge of the ocean floor was limited and would continue to be for another forty years. So that can't be the only reason continental drift was rejected.

One legitimate criticism of continental drift was that Wegener could not come up with a mechanism to explain why and how continents would move. The Americas are large and heavy, so if they are moving away from Europe and

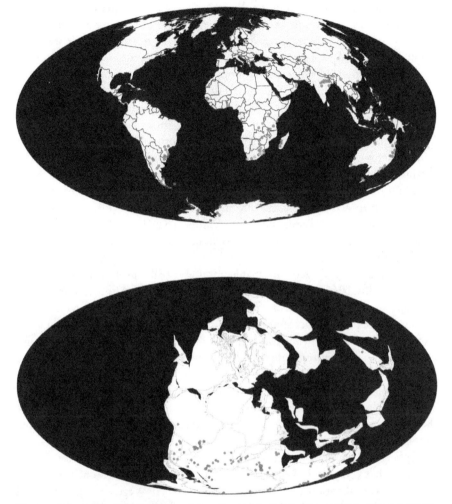

Figure 9.1. The distribution of fossil plants from the genus *Glossopteris*. When occurrences of *Glossopteris* are mapped onto our modern world (top), it implies that the genus has crossed vast oceans. Putting those same occurrences on a map of Pangea (bottom) shows a more sensible distribution. *Glossopteris* once lived in vast forests that covered southern Pangea. The modern distribution is the result of the continents moving, not the forests. *Courtesy of the Paleobiology Database Navigator (paleo biodb.org/navigator).*

Africa, there must be considerable forces involved. Wegener couldn't come up with a source for those forces. Since most of the motion he was describing happened in an east-west direction, he thought it might have something to do with the rotation of the earth, or maybe even the tides,[12] but nothing he could come up with seemed sufficient. Interestingly, this lack of a mechanism is the most commonly offered explanation today as to why geologists rejected Wegener's theory. It is also an explanation that does not hold up to scrutiny.

It can't simply be the case that continental drift was rejected because it lacked a mechanism. Scientists regularly accept the existence of phenomena without knowing their underlying mechanisms. For example, a decade before Wegener proposed continental drift, Einstein proposed that mass interacts with spacetime to create relativistic effects. But to this day, nobody knows why the interaction happens.[13] Biologists accepted Darwin's theory of descent with modification with no idea of how inheritance worked. Even geologists have been known to accept phenomena without explanations. We have known for more than one hundred years now, for example, that the polarity of the earth's magnetic field can occasionally reverse itself, but are still working on exactly why and how that happens.[14]

Furthermore, a plausible mechanism was eventually proposed for continental drift. In his 1945 book *Principles of Physical Geology*, Arthur Holmes, who still supported continental drift after everyone else had dismissed it, wrote, "The movements required to account for the mountain structures are in the same directions as those required for continental drift, and it thus appears that the subcrustal convection currents discussed on pages 408 to 413 may provide the sort of mechanism for which we are looking."[15] The convective motion of the mantle, the same mechanism that Lord Kelvin overlooked in his dating of the earth, was absolutely sufficient to explain the motion of the continents. But by the 1940s, nobody really cared anymore. Continental drift had been relegated to the realms of pseudo-science.

A very bad reason to reject continental drift (or any theory, for that matter) would be that you didn't like the person who proposed it. Sadly, one reason that geologists were skeptical of continental drift was that Wegener, who proposed the theory, had no formal training in geology.[16] This fact was a frequent talking point for the geologists whose theories he was trying to overturn.[17] In 1928, the American Association of Petroleum Geologists (AAPG, one of the largest professional organizations in geology) held a symposium debating Wegener's theory that resulted in a publication called *The Theory of Continental Drift*. The disdain that Wegener's opponents had not just for continental drift but also for Wegener himself practically drips off the page:

- "Facts are facts, and it is from facts that we make our generalizations from the little to the great, and it is wrong for a stranger to the facts he handles to generalize from them to other generalizations."
- "[Wegener writes like] an advocate rather than an impartial investigator."
- "The method of presentation is not scientific."[18]

Rejecting an argument because you don't like or trust the person making the argument is a type of logical fallacy called an *ad hominem* attack. In its structure, it is the reverse of the argument from authority. Instead of saying that something must be right because of the authoritative figure saying it, an *ad hominem* attack ignores the argument and attacks the person making it. Continental drift should have been judged on its merits as a theory, not on the credentials of the person proposing it. "Wegener isn't a geologist" and "Wegener doesn't write like a scientist" are attacks on Wegener himself and are irrelevant to whether or not continents move.

Finally, part of what doomed continental drift is the fact that it was new and different. Nobody had ever seriously considered the possibility that the continents could move before; therefore, continents couldn't move. This is, once again, the appeal to tradition fallacy.[19] The fact that this fallacy was at least partially to blame for geologists rejecting Wegener is front and center in this quote from R. T. Chamberlin, a geologist at the University of Chicago, and a contributor to the AAPG's 1928 volume: "If we are to believe Wegener's hypothesis we must forget everything which has been learned in the last 70 years and start all over again."[20]

A Shift in Perspective

In the wake of the continental drift debate, the 1940s and 1950s were a challenging time for geology. Wegener had raised doubts about theories of geology based on a contracting earth, but nobody had proposed an acceptable alternative yet. There were some *mobilist* geologists out there (as devotees of continental drift had come to be termed), but most geologists were *fixists*, convinced that the position of the continents did not change. In 1958, Charles Hapgood, a professor of history at Keene State College, published a new theory with a unique combination of features that could annoy both mobilists and fixists alike.

In *Earth's Shifting Crust*, Hapgood made the argument that since the earth's crust is rigid, and since it floats on the viscous mantle, the entire crust can move relative to the interior of the earth. This happened periodically and resulted in all of the continents moving *en masse* to new latitudes. During these *crustal shifting* events, portions of the crust could buckle and crumple as they moved, giving us mountains, valleys, and other large-scale topographic features.[21]

Hapgood argued that these sudden shifts were a result of an uneven distribution of ice on earth. He envisioned the crust as being more or less like a car making a sharp turn. If that car was too top-heavy (if, for example, it had too much luggage on the roof), it would tip over. Likewise, too much ice at one pole could cause the entire crust to shift as the earth hurtled along in its curved orbital path.[22] This doesn't work as an analogy for several reasons. To begin with, the weight of ice at the poles is nowhere even close to the weight of the crust.[23] You would be more likely to see a person fall over because they had their hat on crooked than you would be to see the crust shift due to the weight of ice at its top. But more importantly, the earth doesn't have a top. Gravity comes from the center of the earth. Ice would experience the exact same pull of gravity regardless of the latitude at which it sat. However, the lack of a plausible mechanism was only one of the problems with Hapgood's theory.

Like Alfred Wegener, Charles Hapgood had no formal training in geology. He was a historian and a medievalist who had developed his own theory about how the earth might work. But that is pretty much where the similarity ends. Unlike Wegener, Hapgood was perfectly willing to base his theory at least in part on arcane historical conspiracy theories.

In 1513, the Turkish cartographer Piri Reis created one of the first maps of the Atlantic coast of South America. Portions of the map are highly accurate, but some are not. Most notably, the Piri Reis map shows the southernmost coast of South America bending far off to the east. Most historians attributed this to bad data, but there was an alternative hypothesis that Hapgood found fascinating. Hapgood claimed that Piri Reis had made his map by compiling not just new maps of the new world but ancient source maps as well, and that one of these ancient maps showed the continent of Antarctica as it looked under the ice.[24] This, he claimed, was proof that Antarctica had been ice-free within historic times and therefore that it had moved to its current polar position suddenly within the past few thousand years. Hapgood claimed that a sudden shift in the crust could easily explain that equally sudden shift in climate.

Hapgood's theories became increasingly outlandish over the course of his life. He would eventually claim that the source maps that Piri Reis used were made by the citizens of Atlantis.[25] Toward the end of his life, he published a transcript of an interview he conducted with Jesus Christ.[26] Why, then, would scientists devote time and energy to his theories of geology? The answer to this question is because at least one very prominent scientist of the time took Hapgood seriously. *Earth's Shifting Crust* has a foreword written by Albert Einstein. In his foreword, Einstein writes of Hapgood's theory of crustal shifting, "I think that this rather astonishing, even fascinating, idea deserves the serious attention of anyone who concerns himself with the theory of the earth's development."[27] The fact that Einstein was willing to write a foreword for Hapgood illustrates two things. First, like Edmond Halley, Einstein could occasionally have an off day. But more importantly, it shows the state of geological theory in the mid-twentieth century. Contracting earth theory had become suspect, but no one had yet developed a compelling alternative, so the field was ripe for just about anything.

Men's Work, Girl Talk, and a Crack in the World

Prior to World War II, most maps of the world looked about 70 percent blank. That's how much of the world is covered by the oceans. The maps clearly showed the surface of the ocean, but what was happening on the ocean floor was anybody's guess. World War II changed our understanding of the seafloor in two important ways. First, the new technology developed during the war would make high-resolution mapping of the seafloor possible. Second, the need for women in the workforce made it possible for Marie Tharp to become a geologist.

Marie Tharp was born in 1920 and grew up moving frequently. Her father was a soil mapper for the U.S. Department of Agriculture, and the family went where his work took him. When Tharp went to college, it never occurred to her that a woman could also be a geologist, as that was seen as "men's work."[28] In 1943, however, the geology department at the University of Michigan admitted women to its graduate program for the first time, and Tharp was among the first class of students. After working briefly for an oil company in Oklahoma, Tharp went to the Lamont-Doherty Earth Observatory at Columbia University in 1948, where she worked for Maurice Ewing.

Ewing (whom everyone called "Doc") was taking on the rather ambitious project of filling in that missing 70 percent of the map that is hidden by the ocean. During World War II, he had developed a new type of sonar for the navy. It sent out a continuous signal that was reflected off the seafloor to a microphone on the ship's hull. The amount of time that it took for the microphone to detect the reflected signal could be converted into a distance to the seafloor. For the first time ever, a ship at sea could be continuously measuring the depth of the ocean beneath it. Ewing's research group chartered Navy ships to continuously crisscross the Atlantic Ocean taking depth measurements.

Ewing's expeditions collected data from 1947 to 1952. One particular feature of the seafloor that the expeditions were investigating was a shallow area called the Mid-Atlantic Ridge. Naval officers had known for over one hundred years that there was a strange plateau in the center of the North Atlantic,[29] but nobody knew how large it was, what direction it ran, or why it was there. Ewing was determined to solve that mystery and would often recruit young scientists with promises of adventure, shouting, "Young man, would you like to go on an expedition to the Mid-Atlantic Ridge? There are some mountains there, and we don't know which way they run."[30]

In her work with Doc Ewing, Tharp's role was to collate and interpret the data that the ships brought back. She could not be one of Doc's young men on an expedition, as naval regulations prohibited female researchers from joining the voyages.[31] Tharp would receive the data from these expeditions as little more than a long string of numbers and points. It was her job to turn that data into a map of the seafloor. When Tharp put all of that data together into a map, what she found changed our understanding of the ocean and of the earth itself. The Mid-Atlantic Ridge, as it turned out, was a long chain of mountains running north to south down the center of the Atlantic Ocean. However, these mountains did not have a single peak. Instead, each mountain had a V-shaped notch at its apex, indicating that the center of the Mid-Atlantic Ridge was a long valley.

The type of valley that Marie Tharp discovered at the center of the Mid-Atlantic Ridge is called a *rift valley*. It forms due to tension and stretching pulling the earth's crust in a region in two opposite directions at once. In short, Marie Tharp had discovered that there was a crack in the earth running down the middle of the Atlantic Ocean and that the crack was being pulled apart and getting continuously wider over time. When Tharp brought her discovery to the attention of her colleague, Bruce Heezen, he refused to believe it, condescendingly dismissing it as mere "girl talk."[32] Heezen's problem with the implication of Tharp's data was that a widening crack in the ocean floor implied that the crust was moving away from that crack. In an article written in 1999, Tharp quoted Heezen as saying, "It cannot be. It looks too much like continental drift."[33]

By the middle of the 1950s, Tharp and Heezen had amassed enough data to prove beyond a shadow of a doubt that there was a crack not just in the middle of the Atlantic but also surrounding the entire world. The Mid-Atlantic Ridge and its central rift valley are connected to ridges in other oceans. Furthermore, it was seismically active. When a new researcher was hired to make a map of earthquake epicenters in the North Atlantic, those earthquakes followed the valley precisely. The earth was cracked and broken, and the crack was getting larger over time.

Marie Tharp created the first map of the ocean floor and discovered the longest mountain range on earth, the Mid-Ocean Ridge System, but it was Bruce Heezen who got the majority of the credit. The first published article to describe the rift didn't mention Tharp at all.[34] When Heezen published an article in *Scientific American* about the discovery of the rift, Tharp's name appears only in a figure caption.[35] The discovery of the rift may have been the most important discovery in the history of oceanography and one of the most important in the history of geology. When Heezen gave a talk at Princeton University in 1957, a faculty member stood up after the presentation and proclaimed, "Young man, you have shaken the foundations of geology!"[36] That declaration turned out to be prescient. Over the next fifteen years, our understanding of how the earth works would fundamentally change.

Expanding Our Knowledge

The discovery of the rift valley in the center of the Mid-Ocean Ridge was the first step toward developing our modern theory of how the earth works. The next step was determining why the rift is there to begin with, which was accomplished just a few years later. The Mid-Ocean Ridge is where new crust is made. The mountains that comprise the ridge are volcanoes continuously pumping out new rock. That rock (called basalt) is what the entire ocean floor is made of. As the ridge creates new seafloor, all of the old seafloor moves symmetrically away from the ridge. This new theory was called the theory of seafloor spreading and was first proposed by two geologists named Harry Hess and Bob Dietz in the early 1960s.[37]

Seafloor spreading makes the remarkable claim that, because of the existence of the Mid-Atlantic Ridge, the Americas are consistently moving away from Europe and Africa. It's not a noticeable change over a human lifetime. The Atlantic is widening at about four inches per year. But it has been doing so for hundreds of millions of years, and little changes add up. This continuous addition of new material to the outside of the earth posed a challenging question for geologists. How is the earth accommodating all of this new crust? For over a hundred years, geologists worked from the premise that the large-scale structures of the earth's surface were a result of the planet contracting. Suddenly they had to address a whole new possibility. Has the earth actually been expanding through time to accommodate the addition of new crustal material?

The idea of an expanding earth actually predates the theory of continental drift. Its most noteworthy proponent was the Italian geologist Roberto Mantovani who published his own theory of continental motion even before Wegener, in the 1880s. According to Mantovani, the reason that the coastlines of the Americas fit so nicely into the coasts of Europe and Africa is that all the continents were once one large landmass. Mantovani then went one step further. According to him, there were no oceans on the early earth. He believed that the earth's surface had been, until relatively recently, nothing but one big global continent. The modern, separate continents had broken apart, and the ocean basins formed when the earth started expanding.[38]

Nobody really took Mantovani's theory seriously. Even Wegener, who would have had a vested interest in discovering a mechanism for moving the continents, believed that the earth has maintained a constant size over time.[39] But the idea of an expanding earth did make a brief comeback during the 1960s due to two discoveries. The first was Tharp's discovery of an enormous spreading crack down the center of all the world's oceans. The second was that the crust that forms the ocean floor isn't very old.

By the 1960s, the dating of rocks from continental crust had been going on for decades, and people were demonstrating that some parts of the continental crust were billions of years old. When geologists began dating chunks of ocean crust, they couldn't find rock anywhere near that old. There was a pattern to the age of the ocean floor that made perfect sense in the context of seafloor spreading; namely, that the further away you got from the Mid-Ocean Ridge, the older the rock was. But only up to a point. There was no seafloor more than about 200 million years old. This led geologists to briefly reconsider Mantovani's hypothesis. Maybe there is no ocean crust more than 200 million years old because there were no oceans prior to about 200 million years ago.

The primary proponent of Expanding Earth theory in the 1960s was an Australian geologist named S. Warren Carey. His argument was that you could not make a theory of moving continents work on an earth with a constant radius.[40] Expanding Earth theory still appears in the geological literature from time to time.[41] Many adherents believe that they are simply completing the "mobilist revolution" that Wegener began and that the modern theory of plate tectonics is still mired in a fixist paradigm with regards to the earth's size.[42] But the argument that there was no ocean on earth prior to 200 million years ago literally doesn't hold water. To begin with, there is a fossil record of marine life going back billions of years. But more importantly, we have discovered where all that old ocean crust goes and gets destroyed.

Life in the Trenches

If you live in Japan, earthquakes are a regular fact of life. The U.S. Geological Survey has an online searchable database that allows you to see all of the earthquakes that happen anywhere in the world. The database shows that in a typical thirty-day period, Japan experiences anywhere from fifty to one hundred measurable earthquakes.[43] Most of these are so small they might go unnoticed. Some of these are larger. Occasionally these earthquakes are catastrophic. Earthquakes are not the only geological hazard common in Japan. The Japanese Islands are also the site of regular volcanic eruptions, typhoons, and tsunami. Consequently, it is not necessarily surprising that some of the first and best work on determining where earthquakes occur, how seismic waves travel through the crust, and how earthquakes might be related to other geological hazards was done in Japan.

Kiyoo Wadati was born in Nagoya, Japan, in 1902. When he was six years old, while sitting on his grandfather's porch, he felt his first major earthquake.[44] He grew up experiencing regular earthquakes. The most monumental earthquake of his life happened when he was a student at Tokyo University. The Kantō earthquake of 1923 devastated cities across Japan, killed tens of thousands of people, and inspired Wadati to pursue a career in geophysics.[45] He eventually went to work at an observatory in Tokyo working to locate the epicenters of earthquakes as they occurred.

In order to locate the precise point at which an earthquake has occurred, geologists take advantage of the many different kinds of waves that an earthquake produces. Primary waves (P-waves) and secondary waves (S-waves) get their names from the order in which seismometers detect them. P-waves travel faster than S-waves and are therefore detected first. The urban legend that dogs can predict earthquakes comes from the fact that they can actually hear P-waves, which move through an area before the other, more noticeable waves arrive.

The travel speeds of P-waves and S-waves differ; therefore, the amount of time that elapses between the arrival of a P-wave and the arrival of an S-wave is a function of distance. Imagine two cars traveling down the highway at constant but different speeds. If one car is traveling seventy miles an hour and the other is traveling sixty miles an hour, you would be able to tell how far they had been traveling by how much earlier the first car arrives at its destination than the second. Similarly, by measuring how much earlier P-waves are picked up by a seismometer than S-waves, geologists can determine how far from the seismometer the earthquake has occurred. If an earthquake is detected by multiple seismometers, then you can compare the distances between those different seismometers to pinpoint the precise location of the earthquake in a method nearly identical to that used by GPS.

In 1927, Wadati used this method to demonstrate that an earthquake that had shaken Honshu earlier that year had a point of origin 300 km below the surface of the earth.[46] This was beyond unexpected. It was downright problematic. Earthquakes occur when one rigid block of crust moves past another rigid block of crust. But Mohorovičić and others had already demonstrated that the rigid part of the crust is tens of kilometers thick, not hundreds. There shouldn't be anything rigid 300 km under Japan, let alone two rigid things to rub against one another.

In a series of papers over the next decade, Wadati would demonstrate that, in addition to the shallow earthquakes that people knew about and the deep ones he had discovered, there were intermediate ones as well. Furthermore, these earthquakes were not randomly distributed. They occurred along planes dipping increasingly deeper into the earth the further inland you go. Intriguingly, these planes consistently intersected the surface of the earth near deep ocean trenches.

The existence of deep points in the Pacific Ocean was first documented in 1875 by the British navy vessel *HMS Challenger*. As part of a survey of the Pacific Ocean, *Challenger* took regular measurements of water depth. These were simple soundings performed by lowering a weighted rope into the ocean until it touched the bottom. While sailing off the coast of Guam, the *Challenger* made a sounding measurement of over twenty-six thousand feet deep. This spot in the ocean is known today as Challenger Deep, a part of the Marianas Trench.

Like the Mid-Ocean Ridge System, oceanic trenches are dramatic features of our globe that were only recently discovered. At its deepest point, the Marianas Trench is nearly eleven kilometers deep. For comparison, the pinnacle of Mt. Everest is a little less than nine kilometers high. Trenches like the Marianas form a ring along the Pacific coasts of Australia, the Americas, and Asia (including Japan).

What Wadati had discovered, although this would not be recognized until much later, was what happens to the ocean crust at a trench. At trenches, ocean crust sinks into the mantle and is eventually destroyed. The plane of earthquakes that Wadati discovered occurs where a thin slab of ocean crust is being pulled down into the mantle, in a geologic feature called a *subduction zone*. That cold slab of crust cools the surrounding mantle enough to make it rigid, allowing earthquakes to occur far deeper in the earth than one would ordinarily expect. Today this planar region is called the Wadati-Benioff Zone in honor of Wadati and a second seismologist, Hugo Benioff. When oceanic crust sinks deep enough into the mantle, it eventually melts. So long as the amount of crust being destroyed at the trenches roughly balances the amount being created at the ridges, there is no reason to believe that the earth either expands or contracts over time. Contracting earth theory and expanding earth theory were both unnecessary.

Plate Tectonics

Human beings live their lives on the continents, and most rarely get to see more than a few feet below the surface of the ocean. It is therefore unsurprising that we would assume that geology is a primarily continental story. Like geocentricity, a belief that geology is continental feeds into our vanity and is a good example of Bacon's idols of the mind. As it turns out, the continents are for the most part red herrings. The history of the continents is driven primarily by the motion of the ocean.

Most of the rock that makes up the crust of the earth is the ocean floor. The ocean floor is a thin (approximately seven kilometers thick) shell of basalt—a dense blend of silicate minerals rich in iron and magnesium. Basalt is a volcanic rock, and ocean floor basalt erupts out of the volcanoes that form the Mid-Ocean Ridge. With each new eruption, the rock created by previous eruptions spreads away from the ridge. When a sheet of ocean floor gets far enough away from the ridge that spawned it, it can buckle under its own weight like a steel tape measure that has been extended too far, and begin to sink back into the mantle. This results in a deep spot in the ocean called a trench. (See figure 9.2.)

The ocean floor is, therefore, broken into a few dozen coherent sheets of rock surrounded on their various edges by ridges and/or trenches. These sheets of the seafloor are called tectonic plates. They include not just the crust but also the upper, rigid part of the mantle as well. Many of these plates are topped by pure basaltic ocean floor, but some have chunks of lighter material, like granite, embedded in them. These lighter sections of crust float higher in the mantle, stick up higher than the basalt of the ocean floor, and are called continents. Alfred Wegener thought they plowed their way through the ocean floor, but in fact, they are better described as passengers, along for the ride wherever the ocean floor happens to be taking them. From a geological perspective, the centers of plates are relatively dull places. The exciting geology happens at plate boundaries—places where two plates touch one another and are moving relative to one another.

If two adjacent plates are moving away from one another, that's called a *divergent boundary*. The Mid-Ocean Ridge is the classic example of such a boundary. For the most part, all of the excitement of the ridge takes place underwater,

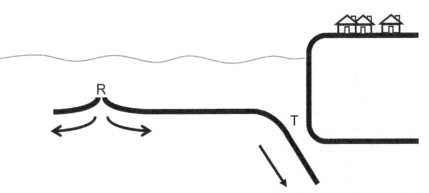

Figure 9.2. The ocean's crust is volcanic in origin. New crust is created at a chain of underwater volcanoes called the Mid-Ocean Ridge (R). The continuous eruption of new crust pushes older crust away from the ridge until it eventually sinks back down into the mantle at deep spots in the ocean called trenches (T). Crust sinking into the trench rubs past the mantle, creating deep earthquakes in the Wadati-Benioff Zone. In the Pacific Ocean, these trenches tend to form along the margins of continents.

but there is one spot where the ridge is actually tall enough to break the surface of the ocean. Iceland is a part of the Mid-Atlantic Ridge. It is a land of earthquakes, volcanoes, hot springs, and geysers all fueled by the fact that western Iceland is moving west, eastern Iceland is moving east, and the part in the middle is being torn apart and stretched out in the process.

When two adjacent plates are moving toward one another, that's called a *convergent boundary*. The islands of Japan and the Pacific coast of Chile are both good examples. At a convergent plate boundary, one plate sinks beneath the other to form a trench in the ocean floor, and a subduction zone going down into the mantle. As the crust sinks back into the earth, it generates deep earthquakes and eventually melts. The melting of the subducting ocean slab typically fuels volcanoes above. Mt. Fuji in Japan and the entire chain of the Andes are examples of volcanoes fueled by the melting of oceanic slabs at subduction zones.

Trenches and subduction zones consume the ocean floor, but continental material is too low in density to sink into the mantle. Therefore, if two continents meet at a convergent boundary, the result is the geologic equivalent of a car crash. The continents collide and make a thickened pile of continental crust along what used to be their coastlines. We call that thickened pile of continental crust a mountain range. For tens of millions of years, India was an island continent being carried northward by the motion of the Indian Plate, until it collided with Asia. The resulting continental collision created the Himalayan Plateau.

There are a few places on earth where plates are moving past one another. These places are called *transvergent boundaries*. Perhaps the best-known example of one of these boundaries is on the west coast of North America, where the Pacific Plate is moving to the northwest relative to the North American Plate. All areas on the west coast of the United States from Cape Mendocino down to the tip of Baja California are technically part of the Pacific Plate. This plate is slowly grinding its way past the rest of the continent, generating massive earthquakes as it goes. Students often ask me if California is going to fall into the sea. It won't. Continental crust doesn't sink. But California may eventually crash into Alaska.

The story of the earth's geography is therefore the story of the birth, life, and death of oceans. New oceans open up, and old oceans close. Against this backdrop, the continents have repeatedly collided to create mountain ranges and have occasionally all come together to make supercontinents. Wegener was right about Pangea, but at 250 million years old, Pangea was only the most recent supercontinent. Earlier supercontinents have had exotic names like Pannotia, Rodinia, and Vaalbara. This dance of continents coming together and shattering repeatedly through geologic time is called the Wilson Cycle, after the Canadian Geologist J. Tuzo Wilson, who provided the final proof for plate tectonics, despite starting out as a skeptic.

The First Mind You Need
to Change Is Your Own

Henrietta Tuzo has two impressive geological namesakes. The first is Mount Tuzo in the Canadian Rockies. Henrietta Tuzo was an accomplished mountaineer and explorer; and in 1906, she became the first person to reach the summit of what was then called Peak Seven.[47] It was to be her crowning achievement as a climber. In 1907, the same year that Peak Seven was renamed in her honor, Tuzo retired from climbing and moved to Ottawa, where she continued to work as an activist for outdoor causes, eventually becoming the president of the Canadian National Parks Association.[48]

Tuzo's second namesake was her son, J. Tuzo Wilson. Born in 1908, Wilson received degrees in geology and geophysics right at the height of the geological debate over continental drift. Several of his professors and mentors were among those who most strongly refuted the possibility of moving continents.[49] As a result, Wilson spent much of his career defending contraction-based models of how the earth worked. In 1950, he published *On the Growth of Continents*, a paper intended to offer a single coherent theory on how mountains formed and the earth in general works. It was based entirely on the premise of a contracting earth, as that was still the prevailing model at the time.[50]

In the 1960s, when some geologists began to consider the possibility of seafloor spreading, Wilson was understandably skeptical. It is a mantra in science that a good theory suggests its own tests, and Wilson wanted to find the test that would definitively demonstrate whether or not the seafloor moves. If the floor of the ocean really does move away from ridges and toward trenches, there should probably be some evidence of that motion somewhere in the millions of square kilometers in between. In 1963, Tuzo Wilson devised a test for seafloor spreading based around the formation of islands.

The Hawaiian Islands sit near the center of the Northern Pacific Ocean and form a chain with the Big Island of Hawaii at the southeast end. As you move northwest through the chain, there are several trends you can see. Moving to the northwest, each island becomes smaller, (less obviously) older, and less volcanically active. Prior to the 1960s, these trends had been attributed to some source of volcanic material moving to the southeast over time along oceanic fault lines.[51]

Wilson offered an alternative explanation. If the floor of the Pacific Ocean was really moving, then all you would need to create a chain of islands like Hawaii would be a single, stationary spot in the mantle that was hotter than its surroundings. Geologists now call this feature a *mantle hotspot*. Wilson argued that as the floor of the ocean moved to the northwest, then at various points in time, each of the islands would have been directly over the hotspot. As each island eventually moved off the hotspot, it would lose its volcanic fuel, stop forming new rock, and begin to decrease in elevation as it cooled. In other words, the further to the northwest an island was, the less volcanic, older, and smaller it would be—exactly the patterns in the Hawaiian Islands.[52]

Wilson did not limit his observations to Hawaii. He pointed out eight other chains of islands in the Pacific that all follow the same patterns as Hawaii.[53] In each case, the islands become smaller, older, and less volcanic along a southeast-to-northwest trajectory—the exact direction that the floor of the Pacific would be moving if it were moving away from the ridge and toward trenches as one single plate.[54] In a second paper that same year, Wilson argued that this same pattern exists in other ocean basins. In every case, chains of volcanic islands demonstrate the motion of the seafloor away from ridges and toward trenches.[55]

In 1965, Tuzo Wilson added an important feature to what would eventually become the theory of plate tectonics. Previously, geologists had believed that trenches and ridges were isolated features of the ocean floor and did not connect to one another. Wilson realized that ridges, trenches, and horizontal faults all connected to one another and that, where they met (at places he called *transforms*), the sense of motion between them could change.[56] By uniting these features, Wilson divided the crust into discrete independently moving pieces. He had added the plates to plate tectonics.

By demonstrating that the crust of the earth moves in discrete, large chunks, Tuzo Wilson changed the way that geologists understood the earth. But the most impressive mind that he changed was his own. When he was fifty years old, Wilson knew that the major features of the earth formed due to a steady contraction of the planet, and that the continents and ocean basins did not actually move or change relative to one another. When he was fifty-five, he had demonstrated to himself and others that almost none of that was true. By the time he was sixty, he was an advocate for a new mobilist theory of geology that his mentors, early colleagues, and even his younger self would have found anathema.

CHAPTER 10

Climate Changes

When Eunice Newton Foote first demonstrated the greenhouse effect, she thought it might be able to explain some of the warmer periods that had occurred in the earth's past. It would be almost another fifty years before anyone really began considering its implications for the future. That's because it was the 1850s and industrial carbon emissions had only barely begun to occur on a large scale. We have known about the link between the carbon dioxide content of the atmosphere and global average temperature for far longer than most people realize, which might explain why so many Americans think that climate change needs more study before government action is taken. How did one of the best-established facts in atmospheric physics become a topic of controversy? Deliberately and recently. This chapter will trace the history of climate change, from our initial realization that the earth's climate changes at all, to the discovery of some of the causes of that change, and finally to the modern fight over what, if anything, to do about the climate change currently being caused by human activity.

The Discovery of Ice Ages

Rocks are reflections of the environment in which they form. The windswept sand of a desert dune field gets preserved as red sandstone—often with the old dune faces still visible in cross-sections. Coal forms primarily in coastal swamps. Today's tropical beaches will one day be incorporated into the rock record as limestone. Knowing these facts makes it easy for geologists to identify warm periods in earth history. Cold periods are a little bit trickier. But there is one mineral that only forms at cold temperatures—ice.

A mineral is any naturally occurring solid substance with a constant chemical formula and a regular arrangement of atoms.[1] Any snowflake checks all these boxes. Ice is not just a mineral; it is one of the most abundant minerals on earth. Like some other minerals and many rocks, ice tells you a lot about the environment it formed in. It was cold. Of course, when temperatures warm up again, the ice melts, but it does leave behind evidence of its existence.

Glaciers are massive and solid, but like the mantle, they are solid things that flow over time. As they traverse the landscape, they bulldoze everything in their path. Even the most massive boulders can be picked up and carried in the flowing ice. Soil and other loose sediment get shoved to the side of the glacier or form a wall at its leading edge. The land underneath the glacier is generally polished smooth by the flow of the ice. If there are rocks embedded in the bottom of the glacier, they will leave gouges in that polished surface showing the direction of glacial flow. When the glacier melts, the water flows away, but those boulders, walls of sediment, and striated surfaces remain behind as evidence that the glaciers had been there. It is these glacial features, and not the ice itself, that allow geologists to see evidence of past periods of extreme cold.

There have been several periods in the earth's history in which glaciers covered far more of the landscape than they do today. What most people typically think of as "the Ice Age," is only the most recent, a fact reflected in the name that most geologists use to refer to it—the last glacial maximum. During that period, glaciers extended so far south that they left boulders and striations behind in what is today Central Park in New York.[2] The walls of sediment that those glaciers left behind were massive things that were pushed right to the edge of the continent. Today those glacial pilings are called Martha's Vineyard and Cape Cod.[3] Geologically speaking, the last glacial maximum was a recent event. Those glaciers only left Central Park about twenty-five thousand years ago.

Conventional wisdom typically attributes the realization that past ice ages have occurred to the Swiss naturalist Louis Agassiz, who first publicly presented evidence of a recent ice age in 1837. But that attribution has always been controversial.[4] Agassiz himself acknowledged that his friend Jean de Charpentier had pointed out the evidence to him that the Rhone Valley was once filled with ice.[5] Some believed that Agassiz had intentionally stolen the idea from the German botanist Karl Schimper.[6] This allegation is somewhat undermined by the fact that the German poet (and avid geologist) Goethe had included the idea in a poem published eight years before Schimper was born.[7] Agassiz and de Charpentier both credited Goethe with being the first to recognize past glacial activity in mountain valleys,[8] but that's not exactly true either.

The simple truth is that no one person can be credited with realizing that glaciers used to be far more extensive than they are today. It was folk knowledge in the Alps long before Agassiz, Schimper, or Goethe came along—a fact that Goethe readily acknowledged.[9] Like the roundness of the earth, the fact that glaciers used to fill alpine valleys was always there for people to see.

Glaciers grow and retreat a little bit each year with the passage of the seasons. As such, anybody who lives and works among modern glaciers can look at the retreating edge of a glacier each spring and see what it has done to the landscape underneath. Glaciers leave very distinctive scars behind on the landscape. If those scars also exist in places where there are no glaciers today, then a good first inference is that glaciers existed there in the past. This logic is an extension of an idea that goes by several names, including Occam's Razor, Uniformity of Process, the Principle of Parsimony, and the K.I.S.S. Principle (Keep it simple, stupid!). They all say the same thing: Don't go making up new explanations if you have one that already works.

Cause and Effect in Climate Change

By the middle of the nineteenth century, geologists and climate scientists had documented that ice ages had occurred in the past. They had also used the distribution of rocks and fossils on the surface of the earth to demonstrate earth had, at times, been warmer in the past. Sometimes those warmer and colder pasts even occurred in the same place. Thirty years after Agassiz demonstrated that Switzerland had been much colder in the past, one of his fellow countrymen, the botanist Oswald Heer, published a fossil history of the region demonstrating that Switzerland had also once been lush and full of palm trees.[10] The rock record was full of evidence for warming and cooling at various points in the past. But what was driving those changes in temperature?

The average temperature on earth is determined by three different factors. The first is sunlight. There are naturally occurring variations in the energy being released by the sun. Also, the fact that the earth's orbit is not perfectly circular means that there are times in the year when we are nearer or farther from the sun. Furthermore, on longer timescales, there are variations in the earth's tilt and orbit due to the wobbling of our planet as it hurtles through space. Each of these factors influences the intensity of sunlight striking the earth's surface, and therefore global average temperature. The second variable affecting global average temperature is the reflectiveness of the planet's surface—what scientists call the earth's *albedo*. Dark surfaces like ocean water absorb sunlight. Shiny surfaces like glacial fields reflect sunlight. The more sunlight the earth absorbs, the warmer it will be. The third variable affecting global average temperature is the content of the atmosphere. What the atmosphere is made of determines its ability to hold and retain the heat from sunlight being reflected off the earth. That doesn't sound like it should be controversial—that the composition of a substance determines its physical properties—but it has become so.

The Greenhouse Effect

Roses are red. Violets are blue. But do you know why these things are true? When we see color, we are seeing the way that light interacts with a particular substance. The light that comes from the sun, or from most lightbulbs, is white light, a combination of every different color of light all mixed together. Roses are red because when light hits them, most of the colors of light are absorbed by the petals of the rose, but the red light is not. The red light reflects off the rose so that the rose appears red.

Roses are red. Violets are blue. But the air looks clear to me and you. You might think that means that the air doesn't absorb or reflect any type of light. However, that's not quite true. Visible light does pass through the air unobstructed, but there are some wavelengths of light that are outside of our ability to see. These include infrared light and ultraviolet light.

In some respects, gas molecules work like wine glasses. Run your finger around the rim of a wine glass at just the right frequency, and it will vibrate and emit a sound. If you expose gas molecules to just the right wavelength of light, they will vibrate and heat up. The visible wavelengths of sunlight don't excite many components of our atmosphere, but the invisible wavelengths of light that reflect off the earth do excite some gases. These are the so-called greenhouse gases.

In 1856, Joseph Henry, the first director of the Smithsonian Institution, read a paper at the annual meeting of the American Association for the Advancement of Science on the relationship between the composition of the atmosphere and its ability to retain heat.[11] This was notable for two reasons. First, it was the first time that the phenomenon, which would eventually be termed "the greenhouse effect," had ever been empirically measured. Second, Henry did not write the paper or perform the experiments himself. The scientist who conducted the experiment was the biologist, inventor, and suffragist Eunice Newton Foote.

In her experiments, Foote put thermometers into glass tubes and then filled the tubes with different gases. She discovered that moist air held and retained heat better than dry air. Most notably, she determined that the strongest effect occurred when she filled a tube with carbon dioxide (or as it was known in those days, carbonic acid). Not only did the tube of carbon dioxide heat more than any other tube, but it also took longer to cool down. Presciently, considering the relative newness of historical geology as a science, Foote remarked:

> An atmosphere of that gas would give to our earth a high temperature; and if as some suppose, at one period of its history the air had mixed with it a larger proportion than at present, an increased temperature from its own action as well as from increased weight must have necessarily resulted.[12]

In the span of two pages, Eunice Newton Foote described the greenhouse effect, identified carbon dioxide as a greenhouse gas, and speculated about the consequences of increased atmospheric carbon dioxide for global climate. Then, her work was largely forgotten. Foote's paper was not included in the proceedings of the conference, most likely because she was an amateur scientist and a woman.[13] Had the presentation not been referenced in another 1857 publication, it may have been lost forever. As it is, modern geologists didn't know it existed until 2011.[14]

The person most frequently credited with discovering the greenhouse properties of carbon dioxide is the Irish chemist John Tyndall. In 1859, Tyndall reported to the Royal Society that he had performed experiments with sealed tubes of various gases and determined that those gases absorb and retain heat differently from one another.[15] This may sound suspicious. Just three years after Foote's findings were reported, but not widely published, a more established male scientist performed a very similar experiment and got his results published by the Royal Society. This seems, however, to be a case of two scientists making the same discovery independently of each other. That happens from time to time in science. For example, Charles Darwin and Alfred Russel Wallace each published theories of natural selection in 1859, and Isaac Newton barely beat Robert Hooke to publishing his theory of universal gravitation.[16] Scientists often refer to these instances of simultaneous discovery as evidence that an idea was "in the air," a metaphor that seems particularly apt for Foote and Tyndall.

There are some key differences between the experiments of Tyndall and Foote that make it unlikely that Tyndall stole anything from Foote. To begin with, Tyndall's apparatus was very different from Foote's. Whereas Foote exposed her glass tubes to direct sunlight, Tyndall used a device that generated infrared radiation.[17] Furthermore, in his initial report to the Royal Society, Tyndall made no reference to one of Foote's most interesting insights—that increased carbon dioxide in the ancient atmosphere might result in a warmer world.[18] His first paper simply described the results of his experiments without offering any practical application.[19] He didn't discuss the possible link between atmospheric content and global temperature until later papers. Finally, since Foote's paper was not widely circulated, there is no evidence that Tyndall, an ocean away, ever found out about it at all.[20]

When Foote and Tyndall first proposed the link between the carbon dioxide content of the atmosphere and global temperature, it was not particularly controversial. In fact, it was incorporated into geology textbooks fairly quickly. One of the most popular and influential geology textbooks of the nineteenth century was J. D. Dana's *Manual of Geology*. Originally published in 1876, it was an attempt to explain discoveries that had been made in geology to an American audience. In his chapter describing the world of the Carboniferous period (approximately 300 million years ago), Dana writes very matter-of-factly:

> Such an atmosphere, containing an excess of carbonic acid as well as of moisture, would have had greater density than the present. . . . It would have occasioned increased heat at the earth's surface, and this would have been one cause of a higher temperature over the globe than the present.[21]

The fact that the atmosphere absorbs and retains heat had been well known even before Foote and Tyndall. After all, if the atmosphere required active heating by sunlight to stay warm, then temperatures would plummet drastically whenever the sun set. Joseph Fourier was the first to point out this fact back in 1827. He was also the first to make the analogy to a greenhouse (or hothouse).[22] Foote and Tyndall demonstrated that it was specific components of the atmosphere that determined the magnitude of the greenhouse effect. In 1896, the Swedish chemist, and future Nobel laureate Svante Arrhenius was able to calculate the specific amount of temperature change associated with changes in carbon dioxide levels.[23] He was also the first person to suggest that burning coal might change the amount of carbon dioxide in the atmosphere, and therefore temperatures. Arrhenius considered this an interesting possibility. Thanks to the rapid acceleration of industrialization throughout the twentieth century, the human species has spent the decades since Arrhenius inadvertently running an experiment to test his hypothesis.

The Case for Anthropogenic Climate Change

For decades now, scientists have been concerned that burning fossil fuels will lead to increases in atmospheric carbon dioxide and therefore to the average temperature on earth as well. This change in global average temperature will result in *anthropogenic climate change*—a human-driven alteration of the large-scale patterns in nature that give society predictability and stability. This argument can be broken down into a series of simple statements: (1) Burning fossil fuels leads to increased carbon dioxide levels. (2) Increased carbon dioxide levels lead to increased temperatures. (3) A warmer world will look different in some fundamental ways from the world of the recent human past. Each of these statements is a testable hypothesis with potential corroborating evidence.

The fact that burning fossil fuels releases carbon dioxide should be relatively uncontroversial. The term *fossil fuels* applies to a broad group of flammable substances, including coal, oil, and natural gas, that are grouped together by the fact that they are the chemical remains of living things. *Burning* is the common term for a process that chemists call *oxidation*. As that alternative term implies, burning involves combining some substance with oxygen (O_2). That's why you can extinguish a fire by cutting off its supply of fresh air.

Of the major types of fossil fuels, coal is the simplest chemically. Coal is carbon (C). All coal has some impurities in it, which is why coal comes in different grades. But what makes all of those substances coal is that they are made mostly of carbon. When you burn coal, you are adding C to O_2. The result is carbon dioxide (CO_2). Oil and natural gas are more chemically complex. Oil in particular is a soup of several different molecules. What all of them have in common with natural gas is that they are molecules called hydrocarbons—composed purely of carbon and hydrogen. They all also have names that end with *–ane*. Some of them will sound familiar like *methane* (natural gas), *propane* (which powers your gas grill), and *octane* (a principal ingredient in gasoline). Oxidize a hydrocarbon, and you get a mix of oxidized hydrogen and oxidized carbon. In other words, you get water vapor and carbon dioxide.

Since the industrial revolution began, and particularly once internal combustion engines became common, the quantity of fossil fuels being burned on earth has very steadily increased. In 1960, the United States emitted an estimated 2.9 million kilotons of carbon dioxide. Today, that number is about 5 million kilotons. This figure doesn't include the emissions of China, Russia, India, or the other two hundred or so countries on earth. All of these combined emissions have significantly changed the content of the atmosphere. In 1960, the atmosphere contained about 300 parts per million (ppm) of carbon dioxide.[24] In other words, for every million molecules in the air, 300 were carbon dioxide. Today the value is closer to 420 ppm[25]—an increase of 40 percent.

Burning fossil fuels necessarily leads to an increase in atmospheric carbon dioxide. Foote, Tyndall, Arrhenius, and others demonstrated the fact that increasing the carbon dioxide content of the atmosphere increases the atmosphere's ability to retain heat. The existence of the greenhouse effect itself is no longer a subject of serious scientific debate. Arguing about the wavelengths of light that a substance absorbs and reflects is (in terms of the underlying physical causes) literally the same as arguing about what color an object is. Carbon dioxide absorbs infrared light and stores the energy from that light as heat. It is an observation as undeniable as the fact that a lemon is yellow.

Beyond laboratory experimentation, there is also data from nature demonstrating the link between carbon dioxide in the atmosphere and global average temperature. According to our best geological evidence, snow and ice have been accumulating on Antarctica for tens of millions of years.[26] As that snow and ice have accumulated on Antarctica, they have trapped bubbles of air within the ice. Climatologists working in Antarctica drill down through the ice and retrieve long cylindrical cores of ice containing these bubbles. By sampling the air in those bubbles, they can analyze the chemical composition of the atmosphere whenever that ancient snow fell. To date, researchers have been able to sample and analyze the content of the atmosphere going back for hundreds of thousands of years.[27] By comparing the content of the ancient atmosphere to the ancient climate record, researchers have been able to demonstrate that global average temperature has been strongly correlated to the greenhouse gas content of the atmosphere for nearly the past half-million years.[28]

People who argue that the greenhouse effect "needs more study" are once again committing the argument from ignorance. The first observation in a lab setting that carbon dioxide increases the ability of the atmosphere to retain heat was done by Eunice Newton Foote in 1856. It has been repeatedly corroborated and understood in more detail in the century and a half since then. Furthermore, measurements of ancient atmospheric content from Antarctic ice cores have demonstrated the strong link between carbon dioxide and temperature over the past half-million years. The greenhouse effect is one of the best-documented correlations in geology. The final question about anthropogenic climate change, then, is this: Will a warmer world look different than a pre-human world?

I have been teaching introductory geology classes for decades. For some topics, I can still use the same notes that I began compiling in the late 1990s. Quartz is still the most abundant mineral on earth. Trilobites are still extinct. Even though Hawaii is moving steadily to the northwest, it is for all intents and purposes in approximately the same place it was last year. But my climate change notes need nearly constant revision. I used to organize my lecture into four sections—underlying theory, data, short-term predictions, and long-term predictions. Today, almost everything that used to be in that third section, short-term predictions, has moved to the data section.

Decreases in ice volume, intensification of hurricanes, widespread drought, changing growing seasons, migration of tropical diseases to higher latitudes, and so many other things that just a generation ago we used to describe as possible consequences are now headlines and history lessons. Yet, according to a recent survey by the Pew Research Center, the percentage of Americans who worry that they will be personally harmed by climate change has gone down by 3 percent in the past six years.[29] Down. This is not the result of a casual misunderstanding. There are millions of dollars spent each year to try to convince the American public that climate change is either not happening or not a problem. The next several sections of this chapter will examine some of the most common arguments used to try to dismiss the idea that anthropogenic climate change is happening.

It's Cold Out

On February 26, 2015, Senator James Inhofe of Oklahoma, who at the time was serving as the chair of the Senate Environment and Public Works Committee, brought a snowball onto the Senate floor. He did it, he said, to remind his colleagues that it was cold out, and therefore that climate change was a myth.[30] This is a very popular tactic among climate change deniers—point out that it is cold in one place to demonstrate that the earth as a whole is not warming. This, of course, is not how an average works. Some variables can decrease in value while the average as a whole goes up. It is possible, for example, for the Dow Jones Industrial to go down on a particular day even if a few of the stocks that comprise that average go up. Likewise, it is perfectly possible for one place on earth to experience a cold snap against a backdrop of temperatures warming on average.

What Senator Inhofe was engaging in is a type of logical fallacy called the cherry-picking fallacy or the fallacy of suppressed evidence. In this type of logical fallacy, you present only the data that supports your argument and either ignore or suppress any data that does not.[31] Climate change deniers commit this fallacy when they concentrate attention on news reports of cold weather in one particular place. News outlets that support climate change denial love to report when there is snow falling on the volcanoes on the big island of Hawaii. Those reports rarely mention that snow falls there all the time. Mauna Kea, the second-highest volcano on Hawaii, has a name that literally means "white mountain" in Hawaiian due to its frequent blanket of snow.[32]

Climate and weather are two different things, which is why there are two different words for them. *Weather* describes the conditions in a particular place and time. *Climate* is the long-term pattern in the weather for a particular place and time. Las Vegas has a desert climate, and it will continue to have a desert climate even if it does rain there next Wednesday. Intentionally confusing *weather* and *climate* is another example of the persuasive definition fallacy. It is at the heart of another common climate change denial argument.

Climate Changes All the Time

The claim that the earth's climate patterns are constantly changing can be true or false depending on exactly what the claimant is arguing. If they are arguing that Chicago being hot in the summertime and cold in the winter is an example of naturally occurring climate change, then they are confusing climate with weather. If, on the other hand, they are arguing that Chicago used to be buried under a glacier but isn't anymore, that's a more legitimate use of the term *climate.* It just still doesn't refute the idea that anthropogenic climate change is occurring.

The geologic record is a record of shifting climates. There are fossil coral reefs in Wisconsin and petrified desert dunes in New England. It is also a record that accumulated over a long period of time. There have been periods in earth history when global temperatures changed with no help from the burning of fossil fuels. The largest temperature increase in the past 60 million years is an event that geologists call the Paleocene-Eocene Thermal Maximum (PETM). During this event, atmospheric and ocean temperatures appear to have risen by 6°C over the course of only about 170,000 years.[33] By geological standards, this is an incredibly rapid change, but it would be nearly undetectable over the course of human lifetimes. Six degrees of temperature rise over 170,000 years comes down to a change of approximately 0.0035°C per century. For comparison, earth surface temperatures increased by approximately 0.6°C during the twentieth century.[34] Put another way, the rate of current warming is approximately 170 times faster than the largest natural events in the geologic record.

Among the geologic history of natural climate changes, the one most likely to provide insights on our current situation might be the temperature drop that occurred over the course of the Carboniferous Period (359–298 million years ago). Between the beginning and the end of the Carboniferous, global average temperatures dropped by about 8°C.[35] The important thing to know about the Carboniferous is why temperatures dropped. As the name implies, the Carboniferous (Latin for "coal-bearing") is when coal formation began on earth.

The Carboniferous was when the first large tree-like plants evolved. Unlike modern trees, which reproduce using seeds, these trees were spore-bearing plants like modern ferns and horsetails. Like their modern relatives, these trees needed to get their spores wet in order to reproduce, and thus lived primarily along coastlines. In other words, the Carboniferous was a time of wide-spread swamps.

Trees spend their lives sucking carbon dioxide out of the atmosphere and converting it into molecules that they use to run their metabolisms and build new cells. Once a tree dies, decomposers usually break its body down and release that carbon back into the atmosphere. If, however, that tree dies in a swamp (as most trees did in the Carboniferous), it could end up getting submerged in stagnant water away from decomposers. When this happens, the carbon in the tree becomes part of the sedimentary rock record, eventually getting concentrated as coal.

The burial of coal in the Carboniferous resulted in a steady drop in the amount of carbon dioxide in the atmosphere. This in turn led to an eight-degree temperature drop worldwide.[36] So yes, climate change can occur naturally. However, even those natural changes demonstrate the power that carbon dioxide has over global average temperature. They also offer us a warning. Burning the world's coal deposits essentially undoes the Carboniferous. It takes the carbon that had been sequestered in the rock and puts it back into the atmosphere. It is therefore only natural to assume that it would undo the Carboniferous temperature drop as well.

But What About . . .

On the night of September 8, 1900, a Category 4 hurricane struck the coast of Galveston, Texas. The winds from this particular storm (which occurred before the practice of naming hurricanes) were blowing at speeds of over 130 miles per hour and were accompanied by a fifteen-foot storm surge. At the time, the highest point in Galveston was less than nine feet above sea level.[37] The 1900 census listed the population of Galveston as 44,116 people.[38] By the morning after the storm, it was approximately 36,000.[39] More than 120 years later, the 1900 Galveston hurricane remains the deadliest single natural disaster in U.S. history.

Climate change deniers often reference Galveston. After all, if hurricanes are becoming more intense due to climate change, then why does the deadliest hurricane in U.S. history predate the automobile? This is a type of argument that sounds valid until you dive into it a little bit deeper. To begin with, there is a difference between being the deadliest hurricane and being the most powerful. The Galveston hurricane was a Category 4 hurricane, with maximum sustained winds between 130 and 156 miles per hour. For comparison, there were four Category 5 hurricanes in 2005 alone.[40]

The Galveston hurricane was particularly deadly in part because of its strength, but more so because it hit a low-lying area with very little warning.[41] It is definitely not evidence that hurricanes were more powerful in the past, and yet it is often referenced that way by climate change deniers. It is a way of subtly changing the subject to something that sounds relevant to the conversation but actually isn't. These off-topic arguments are common in climate debates and can come in two broad types, both of which qualify as logical fallacies.

A strawman fallacy involves arguing against a caricature of a position rather than the position itself.[42] Nobody seriously believes that there were no deadly hurricanes prior to the advent of modern climate change. Therefore, pointing out an event like Galveston doesn't actually contribute substantively to the conversation. The frequency and intensity of hurricanes are limited in part by ocean surface temperatures. As those temperatures rise, more powerful hurricanes will occur more frequently. Pointing out that people were killed by hurricanes in the past doesn't change that fact.

The second type of subject-changing fallacy that often comes up in discussions of climate change is the *tu quoque* fallacy. This fallacy is an attempt to undermine a person's position by pointing out that they are not behaving like a perfect exemplar of that argument.[43] The classic example here would be calling out an actor for using their Oscar acceptance speech to bring attention to the problem of greenhouse gas emissions and then getting into a stretch SUV to be taken to their private jet. Is that behavior hypocritical? Yes. Does it invalidate 150 years of discoveries in atmospheric physics? No.

But I Read a Report by Another Group of Scientists That Said . . .

In 2009, the Nongovernmental International Panel on Climate Change (NIPCC) published a report titled *Climate Change Reconsidered*. This 868-page tome makes some rather surprising assertions: most of the data that supports the model of anthropogenic climate change has been mismanaged; what little warming has occurred in the recent past is primarily driven by causes other than greenhouse gases; and a warm earth would be beneficial to humans, plants, and animals.[44] This all may seem like a shocking reversal. After all, didn't the NIPCC win the Nobel Peace Prize just a few years earlier for working to combat climate change? Actually, no. That was the Intergovernmental Panel on Climate Change (IPCC). The NIPCC is a different group that named itself in a similar way to a legitimate scientific organization in order to cause confusion among the public. The NIPCC used a marketing technique commonly employed by generic breakfast cereals and direct-to-video action-movie knockoffs to make you take them more seriously. This naming practice is not typically used by research institutions looking to debate in good faith.

The NIPCC makes no bones about the fact that its name is very similar to the IPCC. In fact, on its website the NIPCC describes itself as an "independent second opinion" to the findings of mainstream science in general and the IPCC in particular.[45] Notably, there are significant differences between the two organizations. The IPCC was established by the United Nations Environmental Programme and the World Meteorological Organization. It is a panel of scientists from 195 different UN member countries who have convened regularly since 1988 to review the current state of climate research. They are an autonomous body that prepares reports summarizing the current state of our knowledge and making recommendations concerning mitigation and impacts for policymakers around the world.[46] The NIPCC, on the other hand, is mostly the work of one man.

Siegfried "Fred" Singer was born in Vienna in 1924. His family fled Austria during the Nazi rise to power and eventually settled in Ohio when Singer was fourteen. He earned a degree in physics from Princeton in 1944 and then was employed by the U.S. Navy working on underwater mine detection and defense.[47] From the end of WWII until the 1970s, Singer had a significant and impressive career. He designed satellites that, if launched, would have predated Sputnik, and when President Richard Nixon established the Environmental Protection Agency in 1970, Singer was the first deputy administrator for policy.[48]

Despite holding several positions in his life in which he could influence public policy on environmental issues, Singer never really believed that the government had any business regulating the environment.[49] He wrote frequently for mass media espousing *laissez-faire* capitalism and market incentives as the best controls for energy production and environmental protection.[50] In 1990, Singer created the Science and Environmental Policy Project (SEPP), a clearinghouse for contrarian views on science. If you were an industry in danger of regulation in the 1990s, the SEPP had your back. It was the argument from consequences on an industrial scale. Their official positions included not just climate change denial but also denying that ozone depletion was harmful, that sun exposure could lead to skin cancer, or that secondhand smoke caused lung cancer.[51] In 2008, the SEPP organized the NIPCC to concentrate specifically on the fight against climate change, or to be more precise, to concentrate on the fight against the idea of climate change.

If it were done honestly, the work done by the NIPCC would be a welcome contribution to the policy discussion around anthropogenic climate change. But the NIPCC's publications are a cavalcade of logical fallacies, red herrings, and non-sequiturs. For example, *Climate Change Reconsidered* makes the argument that a warmer world might actually lead to improved human health because

> when temperatures rise, death rates fall; when temperatures fall, death rates rise. Bull and Morton (1978) concluded "there is a close association between temperature and death rates from most diseases at all temperatures," and it is "very likely that changes in external temperature cause changes in death rates."[52]

However, Bull and Morton (1978) isn't a climate study. It's a study from a gerontology journal demonstrating that elderly people are more likely to die from non-cancer-related causes in the winter than in the summer.[53] Essentially, the NIPCC's argument in this passage is that if we allow temperatures to rise, then fewer people will have heart attacks shoveling snow. This is probably true, but it is hardly the most pressing issue in the climate debate. This is not an isolated example. The report goes on like this, drawing a rosy picture of our warmer future by misinterpreting hundreds of scientific reports.

NIPCC reports also distort their own importance. One measure of the influence that a scientific paper has is the number of other papers that cite it as a resource. The NIPCC website includes a list of over one hundred different scientific books and articles that have cited *Climate Change Reconsidered*.[54] What they don't tell you is the context in which it is cited. There are many instances of articles citing *Climate Change Reconsidered* as an example of extreme views,[55] articles on the politicization of climate change,[56] and one example in which education researchers used it as an example to help teacher candidates distinguish science from non-science.[57] Frustratingly, this book you are reading might be used in the next iteration of the list to show the ever-increasing influence of *Climate Change Reconsidered*.

Organizations like the NIPCC and the SEPP are particularly problematic because of the feedback that exists between these organizations. One think tank will publish a report throwing doubt on some small detail of a climate study. That report gets cited in another report from a different organization, which is then used as evidence of credibility by the first organization.[58] The reports don't need to contain much in the way of content or merit. It is their existence alone that feeds confusion. If two different groups are telling you two different things about a topic you have only a passing familiarity with, it can be hard to decide whether either of them is right. And that's the whole goal—not to convince you, but to confuse you.

So What If It Is Happening?

We are now to the point in climate science where some questions are settled. The greenhouse properties of carbon dioxide are no longer an opinion or even a theory. They are an observation—a measurement. The same is true for the historic relationship between the greenhouse gas composition of the atmosphere and global average temperature. If these things were not settled, scientists would want to actively debate them. Debate is good for science. It exposes weaknesses in theories and allows scientists to critically examine data and conclusions. There are, in fact, good debates to be had about anthropogenic climate change, but whether climate change is happening or why it's happening aren't among them.

One debate that might be worth having is whether or not we should do anything about it. After all, maybe climate change is the inevitable cost of progress. Going around the world in eighty days used to be science fiction. Now getting to the moon and back only takes a week. Fossil fuels power smartphones, insulin pumps, and countless other things that would have seemed miraculous to Eunice Newton Foote or John Tyndall. Maybe a modern lifestyle is worth the environmental damage. People of good will and honest intent could take either side of that argument; but, if we're going to have that debate, we need to enter into the discussion in good faith. Right now, that is difficult because too many people find it easier to misinform than to win the argument on the merits of their position. Just for the sake of argument, though, let's ask the question of whether or not we should address anthropogenic climate change even if we are not 100 percent sure it is happening.

To address this question, environmental philosophers often turn to a seventeenth-century French mathematician and philosopher named Blaise Pascal.[59] Pascal's final work was the posthumously published *The Pensées*. The fact that Pascal died before it saw publication is particularly ironic because, in one of the most cited sections of the book, Pascal contemplates our final fate. In a construct that has come to be known as Pascal's Wager, Pascal says that all of life comes down to a gamble on the outcome of one question: Does God exist?[60] You can take one of two positions on Pascal's Wager. You can believe that God exists and act accordingly or not, and you can be right or wrong. The combination of possible outcomes can be drawn as a 2×2 grid, as shown in figure 10.1.

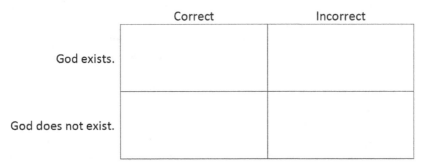

Figure 10.1. According to Pascal, you can either behave as though God exists or behave as though He does not exist. You can either be right or wrong about your assumption.

Let's review what Pascal says happens in each box. If you behave as though God exists and you turn out to be right, that's a great combination. You live a structured and fulfilling life that ends in eternal reward. If you behave as though God exists and you are wrong, that's okay. You don't get the eternal reward, but you do still get a structured life and a sense of purpose. At the end of it all, the lights go out and you cease to exist, so you never even really have to face the disappointment of having been wrong. If you behave as though God does not exist, and are right, even that's a pretty good box. That's the "What happens in Vegas stays in Vegas" box. You can live a life of debauchery and excess with no consequence. Then, there's that last box. If you behave as though God does not exist and you are wrong, you are in big trouble. That box is damnation and eternal torture in the lake of fire.[61] (See figure 10.2.)

	Correct	Incorrect
God exists.	Very Good Outcome	Pretty Good Outcome
God does not exist.	Pretty Good Outcome	**AVOID AT ALL COSTS!!**

Figure 10.2. If you behave as though God does not exist and turn out to be wrong, the consequences are catastrophic. That box should be avoided at all costs.

There is one really good box in the grid, two pretty good boxes in the grid, and one box that you want to avoid at all costs. You have no control over which column you are going to end up in. The existence or non-existence of God isn't up to you. But you can control which row you are in. And that is why Pascal says that a rational person should behave as though God exists, even if they've got some doubts.

What does the grid look like for climate change? Again, we can either behave as though it's a problem or not, and we can be right or not. If we behave as though climate change is a problem and we are right, then not only do we have the opportunity to avert disaster, but also along the way, we get renewable energies, cleaner air, quieter streets, and a lot of other nice things as well. If we behave as though climate change is a problem and we are wrong, we still get the cleaner air, quieter streets, and so on. If we behave as though climate change is not a problem and we are right, then we will keep burning coal and oil for a little while longer. This is a box full of inertia—just the same old world we live in today. But then there's that last box. In that box sits environmental catastrophe.

What the analogy to Pascal's Wager shows is that behaving as though climate change is happening is in our own self-interest.[62] This includes our economic self-interest. By some estimates, the world spends hundreds of billions of dollars every year on climate-related disasters.[63] Even adjusting for inflation, that number has skyrocketed over the past few decades.[64] This doesn't include the cost of reclaiming old mine sites, or the cost in money and lives of mining-related disasters, illnesses, and accidents.

At the same time, the cost of alternative renewable energies has continued to come down. Fossil fuels have powered miracles in the past, but perhaps the most miraculous thing they have done is power the discoveries that are making them obsolete. Thanks to new discoveries, the cost of running the world on solar, wind, and other low/no-emission energy sources is tantalizingly close to the cost of running it on fossil fuels. Fossil fuels have powered the past, but the future of energy is clean and renewable. All climate change denial does is postpone that future. Fighting climate change isn't too expensive. Denying climate change is too expensive.

CHAPTER 11

Our Knowledge Changes

There is an apocryphal story in which a pope asks Michelangelo about his process for sculpting David. In the story, Michelangelo replies that it is actually quite simple: "You take a block of marble and a chisel and you remove everything that is not David." This incident almost certainly never happened, and it is not a particularly good description of the artistic process, but it is often used as a metaphor for the process of scientific discovery. In some respects, science is the systematic exclusion of all the ways the universe does not work. Often, we find the right answer by discarding our old theories in light of new discoveries. There is a crucial difference, however, between the process of sculpting marble and the process of sculpting knowledge: over time, marble dulls a sculptor's tools.

When Thales first predicted his eclipse, we didn't know much about the earth. We knew it was round, but that was about it. Everything else we know about the earth we have learned over time. We learned slowly at first. It was nearly two thousand years later before Copernicus and Galileo were even able to identify where the earth is in space. As our learning increased, however, so did the pace of our learning. This is because, over time, the process of discovery sharpens our tools for new discovery.

The tools that scientists use to understand the world around us fall into several categories. The oldest of them is observation. Posidonius and Eratosthenes used observations of stars and shadows to determine the circumference of the earth with remarkable precision. Ancient astrologers' observations of the retrograde motion of the planets helped Copernicus and Galileo realize that those planets couldn't just be orbiting the earth.

Observation may not seem like a tool that can be honed over time, but each new observation leads scientists to ask new questions and in turn make new observations. The idea that the continents and oceans could move started with Alfred Wegener's observations about coastlines, was revived by Marie Tharp's observation of rift valleys along the Mid-Ocean Ridge, and was corroborated by Tuzo Wilson's observations of Pacific islands. In essence, each new observation prompts us to ask what else we might have missed and inspires new searches. Even misguided questions can prompt important observations. James Hutton was trying to demonstrate an incorrect theory of the earth when he made the observations that helped him realize that the earth was far older than we had thought.

Another important tool of science is coherence. A theory is an explanation for our observations. Like any good story, a good theory ties all of its threads together and resolves them neatly. Copernicus's heliocentric theory is a better one than Ptolemy's geocentric theory in part because it is a simpler story. Copernicus didn't need to rely on epicycles to explain why planets move differently than stars. When Newton developed his law of gravity, it made Copernicus's model even more compelling by demonstrating that those planetary motions could be explained by the same forces that move cannonballs and falling leaves.

The geological career of Nicolas Steno demonstrates the power of a coherent theory to shape our observations and the questions we ask. Steno's theory that glossopetrae were fossilized shark teeth, and not a platonic manifestation, fundamentally changed the way we interpreted fossils. It would enable the observations of Cuvier, Smith, and generations of others, eventually leading to the creation of a worldwide geologic time scale—the framework within which we tell all other geological stories. However, Steno's adherence to a different narrative, the six-day creation of Genesis, kept him from taking part in the scientific revolution that his own work began.

Increasing observational data and improved coherence in our theories, in turn, enable one of the most powerful tools that geologists and all scientists have at their disposal—discussion. Discussing ideas with peers helps scientists to see interpretations they may be missing or holes in their data that they may not recognize. In between sketching his first evolutionary tree and publishing *On the Origin of Species*, Darwin circulated his ideas among his friends for decades, soliciting their opinions and feedback. Copernicus, as well, let his friends read the *Commentariolus* long before he ever published heliocentricity for a broader audience.

Debate and discussion have improved our understanding of several important aspects of the earth. Lord Kelvin and Ernest Rutherford both initially presented their estimates for the age of the earth as public lectures that were open for questions and comments. Eunice Newton Foote's discovery of the greenhouse properties of carbon dioxide was introduced in a similar fashion (if not by Foote herself). Today, thousands of geologists attend annual conferences of organizations like the American Geophysical Union and the Geological Society of America to share the work they are conducting and receive feedback from their peers prior to publication.

The importance of discussion in science underscores the degree to which exclusion has undermined our progress in discovery and learning. Eunice Newton Foote, Marie Tharp, and Inge Lehmann were all dismissed to varying degrees because of their sex. Alfred Wegener's theory was rejected because he was not a geologist. Thomas Jefferson's views on landscape evolution were accused of being too French. While the geological community has made efforts of late to widen the circle of conversation, many voices are still not included in our discussions of the earth. In 2016, underrepresented minorities comprised 31 percent of the U.S. population but only 6 percent of the geoscience doctorates awarded.[1] As representation improves in the future, so will the breadth of our conversations, and inevitably our science.

Keeping some voices out of the conversation certainly makes discussion less effective than it could be. Clinging too strongly to beloved theories of the past can also impede our conversations and our understanding. The Alvarez impact hypothesis for the extinction of the dinosaurs defied conventional wisdom on the pace of geologic change. Wegener's proposition that continents could move required rethinking decades of geological theory. But continents do move (even if Wegener did have the details wrong), and an asteroid did kill the dinosaurs. The fact that we used to believe something else in the past is not a good reason to reject new discoveries.

In the light of new data, it's always a good idea to reconsider your conclusions. This is one of the rules of honest discourse and good-faith discussion. Unfortunately, it also runs contrary to human nature. People don't want to find out, or even worse admit, that they were mistaken. We have a deeply ingrained desire to be correct. Neuroscientists have demonstrated that counterevidence not only fails to sway most people about their beliefs but can also actually make them dig in their heels and cling to those mistaken beliefs even more strongly.[2] This finding about the human mind makes those instances where scientists have been able to accept new theories, whether it was Tuzo Wilson coming around on plate tectonics or Lord Kelvin nodding his acquiescence about the age of the earth, all the more impressive. It also explains how some of Cyrus Teed's followers could continue to preach his divinity and immortality even after his death.

The unfortunate thing about the human desire to be right is that it leaves a door ajar for hucksters and bad actors to try to convince you with false evidence and flawed arguments. If they can convince you to take a position on an issue before you hear all the facts, then you are very likely to stick to that position even after you have heard what should be compelling evidence against it. Scientists have their tools for discovery. People looking to sow confusion and misinformation have their toolkit as well, and discussion is one of their favorite tools.

However, while good scientific discussion is based on a common acceptance of observed data and an attempt to create a coherent explanation for that data, antiscientific discussion is largely based on misdirection and fallacy.

One of the most powerful tools for denying scientific theories is to simply pretend that there is no compelling evidence to support them. This is the argument from ignorance, and it has taken many forms in the history of geological debate. Intelligent Design advocates claiming that life is too complex to be explained by evolution and flat-earthers dismissing experimental results as inexplicably flawed may sound like they are saying different things, but they are both making the same basic argument. They are arguing that their conclusions have never been disproven by refusing to acknowledge the evidence that clearly disproves them.

One tactic that has frequently been employed in the history of arguments about the earth is to subtly change the subject. This is the basis for the slippery slope fallacy, the appeal to consequence, the *ad hominem* attack, and numerous other logical fallacies. These fallacious arguments have been used to undermine geological theories ranging from plate tectonics to natural selection. It is much easier, for example, to argue against Social Darwinism than it is to argue against what Darwin actually said.

Instead of changing the topic, some antiscientific fights are waged by changing the meaning of words. This is the fallacy of persuasive definition. It's the reason that climate change deniers talk so frequently about the weather instead of the climate. It is also behind the frequent dismissal of evolution (or climate change or heliocentricity or radioactivity or whatever else) as "just a theory." In the terminology of science, that dismissal makes no sense. A theory is an explanation. Saying "evolution is an explanation," would actually be a form of agreement, but saying "just a theory" draws on a more common usage of the word to make that explanation seem weak and our understanding look incomplete.

All of these tools used to undermine faith in science are rhetorically powerful but logically weak. They may create doubt, but they cannot create understanding. Understanding is the purview of science. Despite the frequent opposition of nay-sayers, geologists have used the tools at their disposal to increase our understanding of the earth since the days of Thales. Doing so has required geologists to do something that their opponents never seem capable of doing themselves—discarding old theories in favor of new ones when the weight of evidence justifies doing so. Basing our opinions and interpretations on the best and most current data is the epitome of scientific thinking. It's a model for life as well. Refusing to change course in light of new information may be hardwired into our human nature, but it is not a virtue. Likewise, drawing new conclusions based on new data isn't a weakness. It's the process of learning. As a species, it's one of the things we do best and one of the best things we do.

Notes

Introduction

1. Robert N. Proctor, "The History of the Discovery of the Cigarette–Lung Cancer Link."
2. Proctor, "The History of the Discovery."
3. Aristotle, *Aristotle's Posterior Analytics*.
4. Aristotle, *On Sophistical Refutations*.
5. Bradley Dowden, "Fallacies," *Internet Encyclopedia of Philosophy*.

Chapter 1

1. Robert B. Strassler and Andrea L. Purvis, *The Landmark Herodotus: The Histories*.
2. Tom Mandel, "Happy Birthday to Science," *Chicago Sun-Times*.
3. Patricia F. O'Grady, *Thales of Miletus: The Beginnings of Western Science and Philosophy*.
4. O'Grady, *Thales of Miletus*.
5. Isaac Asimov, *Asimov's Biographical Encyclopedia of Science and Technology*.
6. Charles Lyell, *Principles of Geology*.
7. Leila M. Gonzales and Christopher Keane, *Status of the Geoscience Workforce 2011*.
8. Luis W. Alvarez et al., *Extraterrestrial Cause for the Cretaceous-Tertiary Extinction*.
9. Bradley Dowden, "Fallacies," *Internet Encyclopedia of Philosophy*.

Chapter 2

1. Anna McMillan, "Flat-Earth Faithful Flock to Edmonton for International Conference," *CBC News.*

2. Glenn Branch and Craig A. Foster, "Yes, Flat-Earthers Really Do Exist," *Scientific American Blog Network.*

3. Washington Irving, *A History of the Life and Voyages of Christopher Columbus.*

4. Isaac Asimov, *Asimov's Biographical Encyclopedia of Science and Technology.*

5. Thomas Little Heath, *A History of Greek Mathematics.*

6. Heath, *A History of Greek Mathematics.*

7. "Eratosthenes—Biography," *Maths History,* https://mathshistory.st-andrews.ac.uk /Biographies/Eratosthenes.

8. Asimov, *Asimov's Biographical Encyclopedia of Science and Technology.*

9. Asimov, *Asimov's Biographical Encyclopedia of Science and Technology.*

10. Edward Gulbekian, "The Origin and Value of the Stadion Unit Used by Eratosthenes in the Third Century B.C."

11. Donald Engels, "The Length of Eratosthenes' Stade."

12. "NGA—Office of Geomatics," https://earth-info.nga.mil/.

13. John Sellars, *Stoicism.*

14. I. G. Kidd and Ludwig Edelstein, *Posidonius: The Translations of the Fragments.*

15. Alan C. Bowen and Robert B. Todd, *Cleomedes' Lectures on Astronomy: A Translation of The Heavens,* 1st ed.

16. Bowen and Todd, *Cleomedes' Lectures on Astronomy.*

17. Bowen and Todd, *Cleomedes' Lectures on Astronomy.*

18. Bureau International des Poids et Mesures, "Resolution 1 of the 26th CGPM," 2018.

19. "NGA—Office of Geomatics."

20. Violet Moller, *The Map of Knowledge: How Classical Ideas Were Lost and Found.*

21. Ahmad Dallal, *Islam, Science, and the Challenge of History.*

22. Dallal, *Islam, Science, and the Challenge of History.*

23. Samuel Eliot Morison, *Admiral of the Ocean Sea.*

24. Morison, *Admiral of the Ocean Sea.*

25. Claude G. Bowers, *The Spanish Adventures of Washington Irving.*

26. Irving, *A History of the Life and Voyages of Christopher Columbus.*

27. Moller, *The Map of Knowledge.*

28. Jeffrey Burton Russell, *Inventing the Flat Earth: Columbus and Modern Historians.*

29. Russell, *Inventing the Flat Earth.*

30. Brannon Braga, "When Knowledge Conquered Fear," *Cosmos: A Spacetime Odyssey.*

31. Bob Schadewald, "The Plane Truth," *The Plane Truth.*

32. Schadewald, "The Plane Truth."

33. Christine Garwood, *Flat Earth: The History of an Infamous Idea.*

34. Samuel Birley Rowbotham, *Zetetic Astronomy: Earth Not a Globe: An Experimental Inquiry Into the True Figure of the Earth, Proving It a Plane, Without Orbital or Axial Motion, and the Only Known Material World.*

35. Rowbotham, *Zetetic Astronomy.*

36. Schadewald, "The Plane Truth."

37. Schadewald, "The Plane Truth."

38. Rowbotham, *Zetetic Astronomy.*

39. Rowbotham, *Zetetic Astronomy.*

40. Waldemar H. Lehn and Siebren van der Werf, "Atmospheric Refraction: A History."

41. John Hampden, *The Popularity of Error, and the Unpopularity of Truth: Having Special Reference to the Old Copernican and Later Newtonian Theory of the Rotundity and Revolution of the Earth. . . .*

42. Hunter, "Wallace's Woeful Wager."

43. Hunter, "Wallace's Woeful Wager."

Chapter 3

1. National Science Foundation, "The State of U.S. Science and Engineering 2020."

2. Archimedes, "The Sand-Reckoner," in *The Works of Archimedes: Edited in Modern Notation with Introductory Chapters*, ed. Thomas L. Heath.

3. Paul Kunitzsch, "Almagest: Its Reception and Transmission in the Islamic World," in *Encyclopaedia of the History of Science, Technology, and Medicine in Non-Western Cultures*, ed. Helaine Selin.

4. John M. Steele, "A Re-Analysis of the Eclipse Observations in Ptolemy's *Almagest.*"

5. George Huxley, "Eudoxian Topics."

6. Bertrand Russell, *The Problems of Philosophy.*

7. F. H. Bradley, *Essays on Truth and Reality.*

8. William James, *Pragmatism, A New Name for Some Old Ways of Thinking, Popular Lectures on Philosophy.*

9. Andreas Vesalius, *De Humani Corporis Fabrica: Basel, 1543.*

10. Niccolò Tartaglia, *Euclide Megarense philosopho: solo introdutore delle scientie mathematice.*

11. Nicolaus Copernicus and A. M. (Alistair Matheson) Duncan, *On the Revolutions of the Heavenly Spheres/Copernicus; a New Translation from the Latin, with an Introd. and Notes by A. M. Duncan.*

12. Nicolaus Copernicus, *Three Copernican Treatises: The Commentariolus of Copernicus, the Letter against Werner, the Narratio Prima of Rheticus.*

13. Jack Repcheck, *Copernicus' Secret: How the Scientific Revolution Began.*

14. Repcheck, *Copernicus' Secret.*

15. 14th Dalai Lama, *The Universe in a Single Atom: The Convergence of Science and Spirituality.*

16. Jacqueline Leo, "From Bouncer to Pope—21 Fascinating Facts about Pope Francis."

17. Repcheck, *Copernicus' Secret.*

18. E. T. Bell, *The Development of Mathematics.*

19. John Gribbin, *Science: A History.*

20. Copernicus and Duncan, *On the Revolutions of the Heavenly Spheres.*

21. Mark Peterson, "Two Lectures to the Florentine Academy on the Shape, Location and Size of Dante's Inferno by Galileo Galilei, 1588."

22. O. Gingerich et al., *The Eye of Heaven: Ptolemy, Copernicus, Kepler.*

23. G. de Santillana, *The Crime of Galileo.*

24. de Santillana, *The Crime of Galileo.*

25. The Galileo Project, "The Telescope," http://galileo.rice.edu/sci/instruments/telescope.html.

26. Galileo Galilei, *Sidereus Nuncius, or The Sidereal Messenger.*

27. Martin Clutton-Brock and David Topper, "The Plausibility of Galileo's Tidal Theory."

28. "Modern History Sourcebook: Galileo Galilei: Letter to the Grand Duchess Christina of Tuscany, 1615," https://sourcebooks.fordham.edu/mod/galileo-tuscany.asp.

29. The Galileo Project, "Tommaso Caccini," http://galileo.rice.edu/chr/caccini.html.

30. The Galileo Project, "Tommaso Caccini."

31. James Waterworth, *The Canons and Decrees of the Sacred and Oecumenical Council of Trent, Celebrated Under the Sovereign Pontiffs Paul III, Julius III and Pius IV.*

32. Bradley Dowden, "Fallacies," *Internet Encyclopedia of Philosophy.*

33. R. J. Blackwell, *Galileo, Bellarmine, and the Bible.*

34. The Galileo Project, "Pope Urban VIII," http://galileo.rice.edu/gal/urban.html.

35. J. Reston, *Galileo: A Life.*

36. Maffeo Barberini, *Poemata.*

37. G. Galilei et al., *Dialogue Concerning the Two Chief World Systems, Ptolemaic and Copernican,* 2nd rev. ed.

38. D. Harker, *Creating Scientific Controversies: Uncertainty and Bias in Science and Society.*

39. Galilei et al., *Dialogue Concerning the Two Chief World Systems, Ptolemaic and Copernican.*

40. Galilei et al., *Dialogue Concerning the Two Chief World Systems, Ptolemaic and Copernican.*

41. Galilei et al., *Dialogue Concerning the Two Chief World Systems, Ptolemaic and Copernican*; J. Hankinson, *Simplicius: On Aristotle on the Heavens 1.1–4.*

42. Galilei et al., *Dialogue Concerning the Two Chief World Systems, Ptolemaic and Copernican.*

43. S. Drake, *Galileo at Work: His Scientific Biography.*

44. M. A. Finocchiaro et al., *The Galileo Affair: A Documentary History.*

45. Barbara Bieńkowska, "From Negation to Acceptance," in *The Reception of Copernicus' Heliocentric Theory,* ed. Jerzy Dobrzycki.

46. I. Newton and A. Motte, *The Mathematical Principles of Natural Philosophy: Philosophiae Naturalis Principia Mathematica.*

47. Newton and Motte, *The Mathematical Principles of Natural Philosophy.*

48. John L. Helibron, "Censorship of Astronomy in Italy after Galileo," in *The Church and Galileo.*

49. Helibron, "Censorship of Astronomy in Italy after Galileo."

50. "History," *The Vatican Observatory* (blog).

51. Pope John Paul II, "Ai Partecipanti Alla Sessione Plenaria Della Pontificia Accademia Delle Scienze (31 Ottobre 1992)."

52. "Next Generation Science Standards," www.nextgenscience.org.

53. National Science Foundation, "The State of U.S. Science and Engineering 2020."

54. Ray M. Bowen et al., "Science and Engineering Indicators 2014," National Science Foundation.

55. Bowen et al., "Science and Engineering Indicators 2014."

56. Francis Bacon and Joseph Devey, *Novum Organum*.

57. Bacon and Devey, *Novum Organum*.

58. Bill Chappell, "Along with Humans, Who Else Is in the 7 Billion Club."

59. Chappell, "Along with Humans, Who Else Is in the 7 Billion Club."

60. Bacon and Devey, *Novum Organum*.

Chapter 4

1. "Deepest Mine," www.guinnessworldrecords.com/world-records/66169-deepest-mine.

2. John Granlund, "The Carta Marina of Olaus Magnus."

3. Edmond Halley, "An Account of the Cause of the Change of the Variation of the Magnetical Needle. with an Hypothesis of the Structure of the Internal Parts of the Earth."

4. Halley, "An Account of the Cause."

5. Edmond Halley, "V. An Account of the Late Surprizing Appearance of the Lights Seen in the Air, on the Sixth of March Last; with an Attempt to Explain the Principal Phænomena Thereof; as It Was Laid before the Royal Society by Edmund Halley, J. V. D. Savilian Professor of Geom. Oxon, and Reg. Soc. Secr."

6. Ruth S. Freitag, "Hollow-Earth Theories: A List of References: Science Reference Guides," Science Reference Services, Library of Congress.

7. D. A. Griffin, "Hollow and Habitable Within: Symmes's Theory of Earth's Internal Structure and Polar Geography."

8. Robert F. Almy, *J. N. Reynolds: A Brief Biography with Particular Reference to Poe and Symmes.*

9. Lawrence Martin, "Antarctica Discovered by a Connecticut Yankee, Captain Nathaniel Brown Palmer."

10. John Adams, "Fourth Annual Message."

11. Frederick John, "The Yawning Hole."

12. J. N. Reynolds and J. Godsey, *Mocha Dick: Or The White Whale of the Pacific.*

13. William Alfred Hinds, *American Communities and Co-Operative Colonies.*

14. Hinds, *American Communities.*

15. James Naylor et al., *Earth Not a Globe Review Jan–Mar 1897.*

16. Bradley Dowden, "Fallacies," *Internet Encyclopedia of Philosophy.*

17. Christopher Hitchens, *God Is Not Great: How Religion Poisons Everything.*

18. "Koreshan State Park," www.floridastateparks.org/parks-and-trails/koreshan-state-park.

19. Sabrina Stierwalt, "How Deep Is the Deepest Hole in the World?"

20. Isaac Newton and Benjamin Motte, *The Mathematical Principles of Natural Philosophy: Philosophiae Naturalis Principia Mathematica*.

21. John Henry Poynting, *The Mean Density of the Earth*.

22. John Henry Poynting, *The Earth: Its Shape, Size, Weight and Spin*.

23. Robert Mentzer, "How Mason & Dixon Ran Their Line."

24. C. Hutton, "XXXIII. An Account of the Calculations Made from the Survey and Measures Taken at Schehallien, in Order to Ascertain the Mean Density of the Earth."

25. Henry Cavendish, "Experiments to Determine the Density of the Earth."

26. Government Publishing Office, *Astronomical Almanac for the Year 2021*.

27. The Illustrated News of London, "Tea in India."

28. Peter Molnar, "The Structure of Mountain Ranges."

29. I. Cave Brown, "The Venerable Archdeacon Pratt, Archdeacon of Calcutta: A Sketch."

30. Molnar, "The Structure of Mountain Ranges."

31. John Pratt, "On the Attraction of the Himalaya Mountains, and of the Elevated Regions beyond Them, upon the Plumb-Line in India."

32. Molnar, "The Structure of Mountain Ranges."

33. D. Herak and M. Herak, "The Kupa Valley (Croatia) Earthquake of 8 October 1909—100 Years Later."

34. C. M. Jarchow and G. A. Thompson, "The Nature of the Mohorovičić Discontinuity."

35. Jarchow and Thompson, "The Nature of the Mohorovičić Discontinuity."

36. Francis Bacon, *Novum Organum*, ed. Joseph Devey.

37. Molnar, "The Structure of Mountain Ranges."

38. Kerry Lotzoff, "Types of Meteorites."

39. Brush, "Discovery of the Earth's Core."

40. Stephen G. Brush, "Discovery of the Earth's Core."

41. Quentin Williams, "Why Is the Earth's Core So Hot?"

42. "Inge Lehmann—Biography, Facts and Pictures," www.famousscientists.org/inge-lehmann.

43. M. Kölbl-Ebert, "Inge Lehmann's Paper: 'P.'"

44. Kölbl-Ebert, "Inge Lehmann's Paper: 'P.'"

45. Kölbl-Ebert, "Inge Lehmann's Paper: 'P.'"

46. Kölbl-Ebert, "Inge Lehmann's Paper: 'P.'"

47. Bruce A. Buffett, "Earth's Core and the Geodynamo."

48. Buffett, "Earth's Core and the Geodynamo."

49. "Inge Lehmann—Biography, Facts and Pictures."

50. Kölbl-Ebert, "Inge Lehmann's Paper: 'P.'"

51. Kölbl-Ebert, "Inge Lehmann's Paper: 'P.'"

52. American Geophysical Union, "William Bowie Medal."

53. American Geophysical Union, "Inge Lehman Medal."

Chapter 5

1. Martin Brasier, John Cowie, and Michael Taylor, "Decision on the Precambrian-Cambrian Boundary Stratotype."

2. Thomas Browne, *Religio Medici: The Religion of a Doctor*.

3. Browne, *Religio Medici: The Religion of a Doctor*.

4. James Barr, "Why the World Was Created in 4004 B.C.: Archbishop Ussher and Biblical Chronology l."

5. Barr, "Why the World Was Created in 4004 B.C."

6. Barr, "Why the World Was Created in 4004 B.C."

7. Jerome Lawrence and Robert E. Lee, *Inherit the Wind*.

8. Alan Cook, *Edmond Halley: Charting the Heavens and the Seas*.

9. Edmond Halley, "A Short Account of the Cause of the Saltness of the Ocean, and of the Several Lakes That Emit No Rivers; with a Proposal, by Help Thereof, to Discover the Age of the World. Produced before the Royal-Society by Edmund Halley, R. S. Secr."

10. Halley, "A Short Account."

11. Halley, "A Short Account."

12. John Joly, "An Estimate of the Geological Age of the Earth."

13. Dennis Dean, *James Hutton and the History of Geology*.

14. W. Bristow, "Enlightenment."

15. Dean, *James Hutton and the History of Geology*.

16. Stephen Jay Gould, *Time's Arrow, Time's Cycle: Myth and Metaphor in the Discovery of Geological Time*.

17. Dean, *James Hutton and the History of Geology*.

18. James Hutton, *Theory of the Earth: With Proofs and Illustrations*, ed. Sir Archibald Geikie.

19. Hutton, *Theory of the Earth*.

20. Hutton, *Theory of the Earth*.

21. Hutton, *The 1785 Abstract of James Hutton's Theory of the Earth*.

22. Hutton, *Theory of the Earth*.

23. John Playfair, *Illustrations of the Huttonian Theory of the Earth*.

24. Playfair, *Illustrations of the Huttonian Theory of the Earth*.

25. Dean, *James Hutton and the History of Geology*.

26. Playfair, *Illustrations of the Huttonian Theory of the Earth*.

27. Hutton, *Theory of the Earth*.

28. Hutton, *Theory of the Earth*.

29. Immanuel Kant, *Universal Natural History and Theory of the Heavens*, ed. Stanley L. Jaki.

30. David Lindley, *Degrees Kelvin: A Tale of Genius, Invention, and Tragedy*.

31. Lindley, *Degrees Kelvin*.

32. William Thomson, "4. On the Secular Cooling of the Earth."

33. Thomson, "4. On the Secular Cooling of the Earth."

34. Charles Darwin, *On the Origin of Species.*

35. Thomas Henry Huxley, "Anniversary Address of the President."

36. Huxley, "Anniversary Address of the President."

37. Richard R. Nelson, "Physics Envy: Get Over It."

38. Carl Sagan, *The Demon-Haunted World: Science as a Candle in the Dark.*

39. Philip C. England, Peter Molnar, and Frank M. Richter, "Kelvin, Perry and the Age of the Earth."

40. William Thomson, "The 'Doctrine of Uniformity' in Geology Briefly Refuted."

41. England, Molnar, and Richter, "Kelvin, Perry and the Age of the Earth."

42. "All Nobel Prizes in Physics," www.nobelprize.org/prizes/lists/all-nobel-prizes-in-physics.

43. Ernest Rutherford, *Radio-Activity.*

44. Ernest Rutherford, *Radioactive Transformations.*

45. Arthur Stewart Eve, *Rutherford: Being the Life and Letters of the Rt. Hon. Lord Rutherford, O.M.*

46. Eve, *Rutherford.*

47. Eve, *Rutherford.*

48. Frank M. Richter, "Kelvin and the Age of the Earth."

49. John Perry, "The Age of the Earth."

50. Josh Rosenau, "Just How Many Young-Earth Creationists Are There in the U.S.?"

51. *Answers Research Journal,* "About."

52. *Science,* homepage, www.science.org.

Chapter 6

1. Troels Kardel and Paul Maquet, *Nicolaus Steno: Biography and Original Papers of a 17th Century Scientist.*

2. Kardel and Maquet, *Nicolaus Steno.*

3. Kuang-Tai Hsu, "The Path to Steno's Synthesis on the Animal Origin of Glossopetrae."

4. Martin J. S. Rudwick, *The Meaning of Fossils.*

5. Pliny the Elder, *Natural History,* ed. John F. Healey.

6. Hsu, "The Path to Steno's Synthesis."

7. Walter Isaacson, *Leonardo Da Vinci.*

8. Rudwick, *The Meaning of Fossils.*

9. Kardel and Maquet, *Nicolaus Steno.*

10. Rudwick, *The Meaning of Fossils.*

11. Alan H. Cutler, "Nicolaus Steno and the Problem of Deep Time."

12. Kardel and Maquet, *Nicolaus Steno.*

13. Bradley Dowden, "Fallacies," *Internet Encyclopedia of Philosophy.*

14. B. J. Novak, "De-Extinction."

15. David M. Raup, *Extinction: Bad Genes or Bad Luck?,* ed. Stephen Jay Gould.

16. Martin J. S. Rudwick, *Georges Cuvier, Fossil Bones, and Geological Catastrophes: New Translations and Interpretations of the Primary Texts.*

17. Rudwick, *Georges Cuvier.*

18. Rudwick, *Georges Cuvier.*

19. Rudwick, *Georges Cuvier.*

20. Georges Cuvier, "Memoir on the Species of Elephants, Both Living and Fossil."

21. Thomas Jefferson, "A Memoir on the Discovery of Certain Bones of a Quadruped of the Clawed Kind in the Western Parts of Virginia."

22. Clement Clarke Moore, *Observations upon Certain Passages in Mr. Jefferson's Notes on Virginia, Which Appear to Have a Tendency to Subvert Religion and Establish a False Philosophy.*

23. Moore, *Observations.*

24. Thomas Jefferson, *Notes on the State of Virginia.*

25. Moore, *Observations.*

26. Jefferson, *Notes on the State of Virginia.*

27. Moore, *Observations.*

28. Moore, *Observations.*

29. Clement Clarke Moore, "A Visit from St. Nicholas."

30. Stephen M. Rowland, "Thomas Jefferson, Extinction, and the Evolving View of Earth History in the Late Eighteenth and Early Nineteenth Centuries."

31. Silvio A. Bedini and Thomas Jefferson, *Thomas Jefferson: Statesman of Science.*

32. Rowland, "Thomas Jefferson."

33. Jefferson, "A Memoir on the Discovery."

34. Jefferson, "A Memoir on the Discovery."

35. Jefferson, "A Memoir on the Discovery."

36. Rowland, "Thomas Jefferson."

37. Rowland, "Thomas Jefferson."

38. Martin J. S. Rudwick, "The Foundation of the Geological Society of London."

39. Geological Society of London, "Appendix I: Geological Inquiries (1808)," in *The Making of the Geological Society of London.*

40. Rudwick, "The Foundation of the Geological Society of London."

41. John Phillips, *Memoirs of William Smith.*

42. Phillips, *Memoirs of William Smith.*

43. Simon Winchester, *The Map That Changed the World.*

44. Phillips, *Memoirs of William Smith.*

45. Winchester, *The Map That Changed the World.*

46. H. S. Torrens, "Rock Stars: William 'Strata' Smith."

47. Torrens, "Rock Stars: William 'Strata' Smith."

48. Rudwick, "The Foundation of the Geological Society of London."

49. David Hume, *An Enquiry Concerning Human Understanding; [with] A Letter from a Gentleman to His Friend in Edinburgh; [and] An Abstract of a Treatise of Human Nature,* ed. Eric Steinberg.

50. Hume, *An Enquiry Concerning Human Understanding.*

51. Charles Lyell, *The Geological Evidences of the Antiquity of Man: With Remarks on Theories of the Origin of Species by Variation.*

52. Charles Lyell, *Principles of Geology.*

53. Lyell, *Principles of Geology.*

54. Stephen Jay Gould, *Time's Arrow, Time's Cycle: Myth and Metaphor in the Discovery of Geological Time.*

55. Anthony Hallam, *Great Geological Controversies.*

56. Gould, *Time's Arrow.*

57. L. W. Alvarez et al., "Extraterrestrial Cause for the Cretaceous-Tertiary Extinction."

58. Bruce F. Bohor, Peter J. Modreski, and Eugene E. Foord, "Shocked Quartz in the Cretaceous-Tertiary Boundary Clays: Evidence for a Global Distribution."

59. Jennifer A. Kitchell, David L. Clark, and Andrew M. Gombos Jr., "Biological Selectivity of Extinction: A Link between Background and Mass Extinction."

60. Alan R. Hildebrand et al., "Chicxulub Crater: A Possible Cretaceous/Tertiary Boundary Impact Crater on the Yucatán Peninsula, Mexico."

Chapter 7

1. Isaac Newton, *The Mathematical Principles of Natural Philosophy: Philosophiae Naturalis Principia Mathematica,* trans. Andrew Motte.

2. Bradley Dowden, "Fallacies," *Internet Encyclopedia of Philosophy.*

3. Francis Bacon, *Novum Organum,* ed. Joseph Devey.

4. Alvar Ellegård, "The Darwinian Theory and Nineteenth-Century Philosophies of Science."

5. Charles Darwin, *The Correspondence of Charles Darwin: Volume 24, 1876,* ed. Frederick Burkhardt and James A. Secord.

6. Richard R. Yeo, *Defining Science: William Whewell, Natural Knowledge and Public Debate in Early Victorian Britain.*

7. Karl Popper, *The Logic of Scientific Discovery.*

8. Karl Popper, *Conjectures and Refutations: The Growth of Scientific Knowledge.*

9. William Whewell, *The Philosophy of the Inductive Sciences.*

Chapter 8

1. David W. Goldsmith, "On the Origin of Species and the Limits of Science."

2. Nicholas J. Wade, "Erasmus Darwin (1731–1802)."

3. Chris Duffin, Richard T. J. Moody, and Christopher Gardner-Thorpe, *A History of Geology and Medicine.*

4. Ernst Krause, W. S. Dallas, and Charles Darwin, *Erasmus Darwin.*

5. Samuel Taylor Coleridge, *Letters of Samuel Taylor Coleridge,* ed. Ernest Hartley Coleridge.

6. Erasmus Darwin, *The Botanic Garden: A Poem, in Two Parts: Part I. Containing the Economy of Vegetation: Part II. The Loves of the Plants: With Philosophical Notes.*

7. Darwin, *The Botanic Garden.*

8. Krause, Dallas, and Darwin, *Erasmus Darwin.*

9. Erasmus Darwin, *A Plan for the Conduct of Female Education, in Boarding Schools.*

10. Krause, Dallas, and Darwin, *Erasmus Darwin.*

11. Radim Kočandrle and Karel Kleisner, "Evolution Born of Moisture."

12. Erasmus Darwin, *The Temple of Nature.*

13. William Paley, *Natural Theology: Or, Evidences of the Existence and Attributes of the Deity.*

14. Paley, *Natural Theology.*

15. Thomas Aquinas, *Summa Theologica.*

16. "Art. III. Natural Theology; or, Evidences of the Existence and Attributes of the Deity, Collected from the Appearances of Nature," *The Edinburgh Review* 1 (October 1802–January 1803).

17. "Art. I. Natural Theology; or Evidences of the Existence and Attributes of the Deity, Collected from the Appearances of Nature," *The British Critic* 22 (1803).

18. John van Wyhe, "A Biographical Sketch of Charles Darwin's Father Robert Waring Darwin, M.D."

19. Charles Darwin, *The Autobiography of Charles Darwin.*

20. From Charles Darwin, "Darwin Correspondence Project, 'Letter No. 208.'"

21. van Wyhe, "A Biographical Sketch."

22. Darwin, *The Autobiography of Charles Darwin.*

23. Darwin, *The Autobiography of Charles Darwin.*

24. Bill Bryson, *At Home: A Short History of Private Life.*

25. From J. S. Henslow, "Darwin Correspondence Project, 'Letter No. 105.'"

26. Darwin, *The Autobiography of Charles Darwin.*

27. Charles Darwin, *Charles Darwin's Beagle Diary (1831–1836).*

28. Darwin, *Charles Darwin's Beagle Diary (1831–1836).*

29. John van Wyhe, "The Complete Works of Charles Darwin Online."

30. Peter J. Vorzimmer, "The Darwin Reading Notebooks (1838–1860)."

31. Thomas Henry Huxley, "On the Reception of the 'Origin of Species.'"

32. Charles Darwin, *On the Origin of Species.*

33. Darwin, *On the Origin of Species.*

34. Darwin, *On the Origin of Species.*

35. "Evolution, n."

36. Darwin, *On the Origin of Species.*

37. Herbert Spencer, *The Principles of Biology.*

38. To A. R. Wallace, "Darwin Correspondence Project, 'Letter No. 5145.'"

39. Stephen Jay Gould, *Eight Little Piggies: Reflections in Natural History.*

40. G. E. Moore, *Principia Ethica.*

41. Bradley Dowden, "Fallacies," *Internet Encyclopedia of Philosophy.*

42. Kevin P. Lee, "Inherit the Myth: How William Jennings Bryan's Struggle with Social Darwinism and Legal Formalism Demythologize the Scopes Monkey Trial."

43. William Jennings Bryan, "The Prince of Peace."

44. Bryan, "The Prince of Peace."

45. William Jennings Bryan and Mary Baird Bryan, *The Memoirs of William Jennings Bryan.*

46. "Logical Fallacy: Appeal to Consequences," www.fallacyfiles.org/adconseq.html.

47. Bryan and Bryan, *The Memoirs of William Jennings Bryan.*

48. K. K. Bailey, "The Enactment of Tennessee's Antievolution Law."

49. Bryan and Bryan, *The Memoirs of William Jennings Bryan.*

50. Bryan and Bryan, *The Memoirs of William Jennings Bryan.*

51. Bailey, "The Enactment of Tennessee's Antievolution Law."

52. Jerome Lawrence and Robert E. Lee, *Inherit the Wind.*

53. James C. Foster, "Scopes Monkey Trial."

54. Foster, "Scopes Monkey Trial."

55. John T. Scopes and James Presley, *Center of the Storm: Memoirs of John T. Scopes.*

56. Clarence Darrow, *The Story of My Life.*

57. Foster, "Scopes Monkey Trial."

58. Foster, "Scopes Monkey Trial."

59. R. Halliburton, "The Adoption of Arkansas' Anti-Evolution Law."

60. Kristine Bowman, "Epperson v. Arkansas."

61. "Susan Epperson et al., Appellants, v. Arkansas," www.law.cornell.edu/supreme court/text/393/97.

62. "Frequently Asked Questions," *Discovery Institute.*

63. Aquinas, *Summa Theologica.*

64. William Safire, "Neo-Creo."

65. Stephen C. Meyer, "Intelligent Design Is Not Creationism."

66. Renyi Liu and Howard Ochman, "Stepwise Formation of the Bacterial Flagellar System."

67. John A. Boehner, "H.R.1—107th Congress (2001–2002)."

68. Joseph Liu, "The Biology Wars: The Religion, Science and Education Controversy."

69. Dowden, "Fallacies."

70. Darwin, *On the Origin of Species.*

71. Darwin, *On the Origin of Species.*

72. Darwin, *The Autobiography of Charles Darwin.*

73. Darwin, *The Autobiography of Charles Darwin.*

Chapter 9

1. Israel C. White et al., *Final Report Presented to H. Ex. Dr. Lauro Severiano Müller, Minister of Industry, Highways and Public Works.*

2. Anthony Hallam, *Great Geological Controversies.*

3. William Scott-Elliot, *The Lost Lemuria.*

4. H. von Ihering, "Land-Bridges across the Atlantic and Pacific Oceans during the Kainozoic Era."

5. von Ihering, "Land-Bridges."

6. H. S. Williams, "James Dwight Dana and His Work as a Geologist."

7. Anthony Hallam, "Alfred Wegener and the Hypothesis of Continental Drift."

8. Hallam, "Alfred Wegener."

9. Alfred Wegener et al., *The Origin of Continents and Oceans.*

10. W. J. Kious et al., *This Dynamic Earth: The Story of Plate Tectonics.*

11. Wegener et al., *The Origin of Continents and Oceans.*

12. Wegener et al., *The Origin of Continents and Oceans.*

13. Michael Strevens, "No Understanding without Explanation."

14. B. Raphaldini et al., "Geomagnetic Reversals at the Edge of Regularity."

15. Arthur Holmes, *Principles of Physical Geology.*

16. Hallam, "Alfred Wegener."

17. Robert Newman, "American Intransigence."

18. W. A. J. M. van Waterschoot van der Gracht et al., *Theory of Continental Drift; a Symposium on the Origin and Movement of Land Masses, Both Inter-Continental and Intra-Continental, as Proposed by Alfred Wegener.*

19. Bradley Dowden, "Fallacies," *Internet Encyclopedia of Philosophy.*

20. van der Gracht et al., *Theory of Continental Drift.*

21. Charles H. Hapgood and James H. Campbell, *Earth's Shifting Crust: A Key to Some Basic Problems of Earth Science.*

22. Hapgood and Campbell, *Earth's Shifting Crust.*

23. B. J. Davies and N. F. Glasser, "What Is the Global Volume of Land Ice and How Is It Changing?"

24. Hapgood and Campbell, *Earth's Shifting Crust.*

25. Charles H. Hapgood, *Maps of the Ancient Sea Kings: Evidence of Advanced Civilization in the Ice Age.*

26. Charles H. Hapgood and Elwood Babbitt, *Talks with Christ and His Teachers: Through the Psychic Gift of Elwood Babbitt.*

27. Hapgood and Campbell, *Earth's Shifting Crust.*

28. Marie Tharp, "Connect the Dots: Mapping the Seafloor and Discovering the Mid-Ocean Ridge."

29. Tharp, "Connect the Dots."

30. Tharp, "Connect the Dots."

31. Erin Blakemore, "Seeing Is Believing."

32. Tharp, "Connect the Dots."

33. Tharp, "Connect the Dots."

34. M. Ewing and Bruce C. Heezen, "Oceanographic Research Programs of the Lamont Geological Observatory."

35. Bruce C. Heezen, "The Rift in the Ocean Floor."

36. Tharp, "Connect the Dots."

37. Harry Hess, *Evolution of Ocean Basins*; Bob Dietz, "Continent and Ocean Basin Evolution by Spreading of the Sea Floor."

38. Roberto Mantovani, "L'antarctide."

39. Wegener et al., *The Origin of Continents and Oceans.*

40. Samuel Warren Carey, *The Expanding Earth.*

41. Dennis McCarthy, "Biogeographical and Geological Evidence for a Smaller, Completely-Enclosed Pacific Basin in the Late Cretaceous."

42. Martin Pickford, "Earth Expansion and Plate Tectonics: Historical Review, Comparison and Discussion."

43. U.S. Geological Survey, "Latest Earthquakes."

44. Kiyoo Wadati, "Born in a Country of Earthquakes."

45. Wadati, "Born in a Country of Earthquakes."

46. Yasumoto Suzuki, "Kiyoo Wadati and the Path to the Discovery of the Intermediate-Deep Earthquake Zone."

47. Susan Markham, "Celebrating Women in the Parks: From Goddesses to Ministers of the Crown."

48. Markham, "Celebrating Women in the Parks."

49. Linda Hall Library, "Scientist of the Day—Tuzo Wilson," www.lindahall.org/tuzo-wilson.

50. J. Tuzo Wilson, "On the Growth of Continents."

51. J. Tuzo Wilson, "A Possible Origin of the Hawaiian Islands."

52. Wilson, "A Possible Origin of the Hawaiian Islands."

53. Wilson, "A Possible Origin of the Hawaiian Islands."

54. Wilson, "A Possible Origin of the Hawaiian Islands."

55. J. Tuzo Wilson, "Evidence from Islands on the Spreading of Ocean Floors."

56. J. Tuzo Wilson, "A New Class of Faults and Their Bearing on Continental Drift."

Chapter 10

1. Dexter Perkins, "Mineralogy—Free Textbook for College-Level Mineralogy Courses."

2. NYC Parks, "Hot Rocks: A Geological History of New York City Parks," www.nycgovparks.org/about/history/geology.

3. Robert N. Oldale, "Glacial Cape Cod, Geologic History of Cape Cod by Robert N. Oldale."

4. E. P. Evans, *The Authorship of the Glacial Theory.*

5. G. F. Wright, "Agassiz and the Ice Age."

6. Evans, *The Authorship of the Glacial Theory.*

7. Dorothy Cameron, "Early Discoverers XXII: Goethe—Discoverer of the Ice Age."

8. Cameron, "Early Discoverers XXII."

9. Matthew H. Birkhold, "Measuring Ice: How Swiss Peasants Discovered the Ice Age."

10. Oswald Heer, *Die Urwelt Der Schweiz.*

11. Eunice Newton Foote, "Circumstances Affecting the Heat of the Sun's Rays."

12. Foote, "Circumstances Affecting the Heat of the Sun's Rays."

13. R. Jackson, "Eunice Foote, John Tyndall and a Question of Priority."

14. Raymond P. Sorenson, "Eunice Foote's Pioneering Research on CO2 and Climate Warming; #70092 (2011)."

15. John Tyndall, "VII. Note on the Transmission of Radiant Heat through Gaseous Bodies."

16. Louise Diehl Patterson, "Hooke's Gravitation Theory and Its Influence on Newton. II."

17. Jackson, "Eunice Foote, John Tyndall and a Question of Priority."

18. Tyndall, "VII. Note on the Transmission."

19. Tyndall, "VII. Note on the Transmission."

20. Jackson, "Eunice Foote, John Tyndall and a Question of Priority."

21. J. D. Dana, *Manual of Geology: Treating of the Principles of the Science, with Special Reference to American Geological History.*

22. M. Fourier, "Mémoire Sur Les Températures Du Globe Terrestre et Des Espaces Planétaires."

23. Svante Arrhenius, "On the Influence of Carbonic Acid in the Air upon the Temperature of the Ground."

24. Theo Stein, "Carbon Dioxide Peaks near 420 Parts per Million at Mauna Loa Observatory."

25. Stein, "Carbon Dioxide Peaks."

26. W. U. Ehrmann and A. Mackensen, "Sedimentological Evidence for the Formation of an East Antarctic Ice Sheet in Eocene/Oligocene Time."

27. J.-M. Barnola et al., "Historical Carbon Dioxide Record from the Vostok Ice Core."

28. J. R. Petit et al., "Climate and Atmospheric History of the Past 420,000 Years from the Vostok Ice Core, Antarctica."

29. James Bell et al., "In Response to Climate Change, Citizens in Advanced Economies Are Willing to Alter How They Live and Work."

30. Ted Barrett, "Inhofe Brings Snowball on Senate Floor as Evidence Globe Is Not Warming | CNN Politics."

31. Bradley Dowden, "Fallacies," *Internet Encyclopedia of Philosophy.*

32. Doug Herman, "The Heart of the Hawaiian Peoples' Arguments Against the Telescope on Mauna Kea."

33. U. Röhl et al., "On the Duration of the Paleocene-Eocene Thermal Maximum (PETM)."

34. S. Tett et al., "Causes of Twentieth-Century Temperature Change Near the Earth's Surface."

35. Steven M. Stanley, *Earth System History.*

36. Stanley, *Earth System History.*

37. Ron Trumbla and Curtis Carey, "National Weather Service Commemorates 1900 Galveston Hurricane."

38. "Galveston County Profile," https://txcip.org/tac/census/profile.php?FIPS=48167.

39. Trumbla and Carey, "National Weather Service Commemorates."

40. National Hurricane Center, "NHC Data Archive," www.nhc.noaa.gov/data.

41. Trumbla and Carey, "National Weather Service Commemorates."

42. Dowden, "Fallacies."

43. Dowden, "Fallacies."

44. Craig D. Idso and S. Fred Singer, *Climate Change Reconsidered.*

45. "About the NIPCC," climatechangereconsidered.org/about-the-nipcc.

46. "History of the IPCC," www.ipcc.ch/about/history.

47. Rachel White Scheuering, "S. Fred Singer."

48. White Scheuering, "S. Fred Singer."

49. White Scheuering, "S. Fred Singer."

50. S. Fred Singer, *Free Market Energy: The Way to Benefit Consumers.*

51. Science and Environmental Policy Project, "Mission Statement," www.sepp.org /about-us.cfm.

52. Idso and Singer, *Climate Change Reconsidered.*

53. G. M. Bull and J. Morton, "Environment, Temperature and Death Rates."

54. NIPCC, "Academic References to Climate Change Reconsidered."

55. B. W. Abbott et al., "Biomass Offsets Little or None of Permafrost Carbon Release from Soils, Streams, and Wildfire."

56. R. E. Dunlap and P. J. Jacques, "Climate Change Denial Books and Conservative Think Tanks: Exploring the Connection."

57. R. Beach et al., "Fostering Preservice and In-Service ELA Teacher's Use of Digital Practices for Addressing Climate Change."

58. Dunlap and Jacques, "Climate Change Denial Books."

59. D. W. Orr, "Pascal's Wager and Economics in a Hotter Time."

60. Blaise Pascal, *The Pensées.*

61. Pascal, *The Pensées.*

62. Orr, "Pascal's Wager and Economics in a Hotter Time."

63. Munich RE, "Record Hurricane Season and Major Wildfires—the Natural Disaster Figures for 2020."

64. National Centers for Environmental Information, "Billion-Dollar Weather and Climate Disasters."

Chapter 11

1. Rachel E. Bernard and Emily H. G. Cooperdock, "No Progress on Diversity in 40 Years."

2. Jonas T. Kaplan, Sarah I. Gimbel, and Sam Harris, "Neural Correlates of Maintaining One's Political Beliefs in the Face of Counterevidence."

Index

Note: Pages in *italics* refer to exhibits.